Yearbook of the Artificial

Yearbook of the Artificial Volume 2

Edited by
Massimo Negrotti

In the different areas of the artificial – defined as the attempt to reproduce natural objects or processes by means of the available technology – researchers and designers often work within their disciplines with little or no information about research being carried out in other areas. Bioengineers, for instance, know little about the problems of Artificial Intelligence researchers, while roboticians often neglect the efforts of medical engineering, etc. The Yearbook of the Artificial discusses current theories, projects and models from a wide range of areas without going into too many technical details. The main assumption is that there are common logical and methodological problems for all disciplines that are concerned with the reproduction of natural objects and processes. The Yearbook provides a forum for all of them, and covers areas such as bioengineering, robotics, A.I., artificial neural networks, artificial life, multimedia, the history of technology, communications, art and music.

In submitting a paper to the *Yearbook of the Artificial*, you grant us permission to publish that paper after a peer review. We ask that your paper be between 20 000 and 24 000 characters, and pertinent to the topics mentioned above. We also require an abstract of no more than 150 words, including a list of keywords. Papers should be submitted in triplicate, typewritten, double-spaced, on one side of the paper. Please include a copy on a floppy disk in a commonly used word processor format.

Only original papers in English should be sent to:

Dr. Domenico Donnanno
dodo@soc.uniurb.it
IMES-LCA
Via Saffi, 15
University of Urbino
I-61029 Urbino (Pu) Italy

Yearbook *of* the Artificial

Nature, Culture & Technology

Vol. 2

Models in Contemporary Sciences

Edited by Massimo Negrotti

In Memory of
Prof. Martino Rizzotti

PETER LANG
Bern · Berlin · Bruxelles · Frankfurt a.M. · New York · Oxford · Wien

Bibliographical information published by Die Deutsche Bibliothek
Die Deutsche Bibliothek lists this publication in the Deutsche Nationalbibliografie; detailed bibliographical information is available on the Internet at ‹http://dnb.ddb.de›.

British Library and Library of Congress Cataloguing-in-Publication Data:
A catalogue record for this book is available from *The British Library*, Great Britain, and from *The Library of Congress*, USA.

Il presente volume è pubblicato nell'ambito della Ricerca 'Modelli dell'utente nelle società avanzate', finanziata dal M.I.U.R. e dall'Università degli Studi di Urbino 'Carlo Bo', COFIN 2002, Prot. n. 2002143319.

This volume is published as part of the research project 'Models of the User in Advanced Societies', co-financed by the Italian Ministry of University Teaching and Scientific Research (M.I.U.R.) and by the University of Urbino 'Carlo Bo', COFIN 2002, Prot. n. 2002143319.

Cover image: Massimo Negrotti
Cover design: Thomas Jaberg, Peter Lang AG

ISSN 1660-1084
ISBN 3-03910-124-2
US-ISBN 0-8204-6864-9

Hochfeldstrasse 32, Postfach 746, CH-3000 Bern 9, Switzerland
info@peterlang.com, www.peterlang.com, www.peterlang.net

Scientific Committee

Table of Contents

ROBERT FRANCK
Introduction. Modeling is the key to science 9

MASSIMO NEGROTTI
Models Thinking 23

DANILA BERTASIO
Representation and self-representation:
the concept of model in the arts 35

GIOVANNI BONIOLO
Theories and models: really old hat? 61

MARIO BORILLO
Technological risks or risks of reason?
A reflection on cognitive technologies 87

ALESSANDRO BRUSCHI
Logical models and the logical method 105

LUCIANO CELI
White bears and washing-machines:
models (and theories) of concepts 137

MARIELLA COMBI
Cultural models in anthropology 163

LUCIANO FLORIDI AND JEFF W. SANDERS
The Method of Abstraction 177

GIUSEPPE LANZAVECCHIA
Models in science are unnecessary but useful 221

ROLAND MÜLLER
Model: The history of the concept and of its use in science 239

GIUSEPPE PADOVANI
An Implicit Model in Sociology: Raymond Boudon 263

ROBERTO POLI
Ontological Models 289

ANTONINO PORRELLO
Theory of the Simulative Models (TMS) 305

FUMIHIKO SATOFUKA
Paradigm shift of Kuhnian model to post-kuhnian one in the history of science, technology and art in Modern Japan 333

ANDREA SORMANO
Max Weber's Methodological Individualism as 'Sociological Grammar' 339

Robert Franck

Introduction. Modeling is the key to science

Abstract:

In this introductory chapter I want to draw attention to the importance of modeling activity in scientific work. Models are not just a very valuable aid to research. They are actually able to represent the structure of things. In this way, modeling is the essential procedural step in doing science. Those who emphasized observation, experimental controls, and the close relations that exist between science and technology were not wrong, but these emphases have long obscured the real aim of science: to proceed from phenomena upward to their principles, that is, to the structures which govern and explain phenomena.

To begin, I will recall that models represent the structure of systems; the system can be real or imaginary. Then I will show that science seeks to model three different kinds of structures when it studies an empirical system. The result is that scientific models have different roles, depending on whether they are intended to represent imaginary systems, or one of the three structures of an empirical system. I conclude with a defense of the fruitfulness of artificial models.

Keywords:
system, isomorphic, realism, mechanism, properties, empirical models, formal models, artificial models

1 Models represent systems

There are different kinds of scientific models: mathematical models, conceptual models, statistical models, geometrical figures, diagrams, geographical maps, computer modeling, neuronal models, etc. The great number of things which get called models creates some confusion, and we may get the feeling that a 'model' is just another name for a catch-all. In reality, though, all the types of representations that we call models have something in common: they represent systems. A system is a set formed from subsets, relations between these subsets, and relations between the subsets and the set they belong to. A system may be linked to an environment, may indeed be a subset of an environment. Two components are thus enough to make a model: representations of relations, and representations of sets or subsets.

The network of relations between subsets in a system and the network of relations between subsets in another system are not the same. The network that belongs to a particular system is called the structure of that system. The structure of the system is that which the model seeks to represent. The structure of a system is more than often complex, and the model only represents then part of it, perhaps only a very small part.

2 The fictive character of scientific models

It is common to emphasize the fictive nature of scientific models. Many proposed models represent imaginary systems. These can play an exploratory role, provide heuristic support, enable us to test hypotheses, or illustrate arguments for didactic purposes. Models can also aid in the conception of technical arrangements that allow us to act on phenomena. But even when a model purports to represent a real system, it appears to be something fictive in nature. For the model offers us a very incomplete representation of the given system, one in which only the structure of the given system is represented (and usually only a small part of that). Moreover, there is often not even any perceptible resemblance between the model and the part of the system that is supposed to be represented. All this leads us to think that models can never be anything but mental constructions, that modeling is not intended to reflect reality, and that models are good insofar they are useful. This point of view is widely shared today. In my opinion, it's better to give it up.

3 The realism of models

The ability mathematical representations have, to teach us about reality, is amazing. How is it that the representation of reality can be accomplished without any dependence on perceptible resemblance to the things? This is not only true for mathematics, but also for ordinary language and in fact for the majority of our representations. Where does the power of scientific models – our particular interest here – to represent the world come from, if not from resemblance? We get a part of the answer when we pay attention to the fact that a model is the representation of a system, whose structure is what is represented. The part of perceptible phenomena which is represented by a model does not include the sensations that phenomena make us feel, but only their structure (or some substructure of that). Resemblance is not helpful in order for the model to conform to the structure to be represented, but only *isomorphism*. There is isomorphism when the specifications of the model correspond to the specifications of the structure which is represented. For example, every subset of the model, and every relation which connects them, must have corresponding to it a subset and relation in the represented structure, and the subsets and relations must have the same specifications.

We thus see, at the outset, that the absence of perceptible resemblance between models and reality is quite compatible with the realism of these models. It is sufficient to recognize that reality itself is not limited to our sensations, or to what is perceptible in the phenomena (reality is not limited to 'sense data'). The structure of phenomena is just as real as the sensations we get from them. As for the constructed character of scientific models, this is also compatible with realism, since the only thing that counts when deciding if a model is realistic or not, is that it should be isomorphic with the structure to be represented. This isomorphism is rarely complete, of course, since the model emphasizes some specifications of the system and neglects others.

4 Three conceptions of scientific activity

Does scientific activity consist in describing phenomena and obtaining accurate representations of them with the help of models? Alongside this conception of the work of the scientist, a conception quite current today which draws inspiration from empiricism – though empiricism is richer and more subtle than the current conception – there is another, also a current position, which draws inspiration from operationalism. For this latter, science is akin to technology, both being attempts to model material and intellectual arrangements in order to be able to act on phenomena through them. For my part, I should like to insist on the fact that scientific activity – and thus scientific modeling also – is not limited to the description of phenomena (including the description of their observable structures[1]), nor to technology. Science also seeks to discover the structures, underlying phenomena, which govern the phenomena.

The three conceptions I have mentioned each reflect part of the work of the scientist – and thus part of the work of modeling as well – and so none of the three is false. It is desirable, I think, to see them as complementary rather than to set up in opposition to any one of them. The description of phenomena and the elaboration of technical capabilities are broad avenues of scientific progress. But the task at hand, which consists in searching for the underlying structures which govern and thus explain phenomena[2], should not be obscured by emphasis on observation or technical know-how.

Someone who thinks that science has no other task than to furnish accurate representations of phenomena has, paradoxically, good reasons for thinking that no scientific model can represent reality. For sooner or later, he becomes aware that there are multiple points of view on phenomena. He ought to select the subsets and relations which will play a role in his model from among many more, while the great-

1 Structures are never *sense data*, but observable structures do exist. For example, through observation we find out about statistical correlations, although the correlations are not sensations.

2 For an in-depth examination of the ways in which structures determine phenomena, and determine them in non-causal ways, see R. Franck 2003 pp. 187-194.

est part of the phenomenon which he is trying to describe will be left to one side. How can one in this position not conclude that our models, influenced as they are by our own preferences and preoccupations, and involving our deliberate turning away from thousands of possible points of view and observations, are not simply the product of one's own thinking, not the accurate representation of some reality? The same conclusion arises for the one who identifies science with technology. From this viewpoint, the efficiency of the technology worked out determines whether or not a model is good. In both cases usefulness is held up as the only criterion for evaluating models.

The question of the realism of scientific models and their role in contemporary science takes on a different aspect when we admit that science also has the job of discovering the structures which underlie phenomena, and which govern them and furnish explanations for them.

5 Modeling the mechanisms of a system vs. modeling its properties[3]

In medicine we may find ourselves ignorant regarding the process within the organism through which a particular remedy acts on an illness. Nevertheless we may observe in systematic fashion the effects which the remedy produces in patients who have different illnesses, who are of different ages, etc. Statistics are a powerful tool for carrying out such observations. Now, finding out how the human organism reacts to a given remedy under different conditions is tantamount to discovering certain properties of that organism. If the doctor knows about these properties, he or she can decide which remedy is proper for the given patient, according, e.g., to the patient's age, or to his other illnesses, etc. Traditional medicine worked just this way, although today doctors take more care than formerly to observe the effects produced by a particular remedy in a systematic way. In the 19th Century, such a practice in medicine was termed 'empirical' to distin-

3 Section 5 here is adapted from R. Franck 2002, pp. 90-92.

guish it from the 'scientific' medicine which was in its infancy. A new form of medicine took responsibility, unlike its predecessor, for explaining the effects produced by medical remedies, and also attempted to explain the effect of pathogens on the organism. The method was to discover intra-organic mechanisms through which causes or inputs (pathogens and remedies) produced their effects or outputs (the appearance or disappearance of symptoms). Following the distinction between empirical and scientific medicine, the discovery of the properties of the human organism, thanks to systematic observation of effects produced by pathogens and by remedies – including the statistical tools which such observation employs – is sub-scientific (C. Bernard, 1966, p. 297-98). From this point of view, science only begins when we ask ourselves about the mechanisms which may explain observed properties, that is, mechanisms which explain why causes produce the effects they produce within the system under examination (in this case, intra-organic mechanisms).

This example from the developing practice of medicine in the 19th century is very suggestive. But the conception of science it illustrates can be applied to any discipline: are we satisfied when we have modeled the properties of a system? Or are we intent on modeling the mechanism within the system, which would allow us to explain its properties?

We may represent to ourselves the distinction between mechanisms within the system and properties generated by it with the help of the metaphor of the black box.

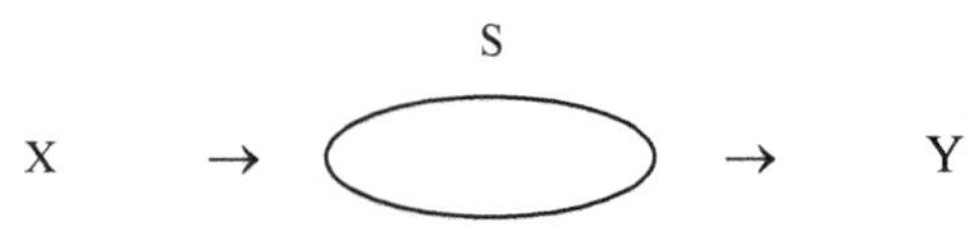

Fig. 1. The properties of a system defined according to a pairing of inputs and outputs

The mechanisms which generate the properties of the system (S) are inside the box. As long as we don't know what's inside the box, we call it 'black'. The properties of the system are outside the box. These are the outputs (Y) which issue from the box when certain inputs (X) occur. No input, no output. We must be careful not to try to define the properties of the system by its outputs alone. A property of the system is that particular kind of output which the system can generate when it receives a particular kind of input. Another system might generate a different output in response to the identical input. And the same system may generate a different output when it receives a different input[4]. The properties of the system must, therefore, be defined according to a pairing of inputs and outputs (X,Y) which is operated by the system. In short, let us define properties as the specific effects that may be produced by the system in interaction with a definite environment.

The metaphor of the black box (which is undoubtedly an imaginary system!) helps us understand the advantage of differentiating the phenomena we observe in this way: some belong to the system's mechanism, while others belong to the properties of the system that are generated by the mechanism. If we don't mark this distinction, we risk confusing a mechanism with its properties in our model when we try to model a series of phenomena. We even risk mixing up within our model fragments of different, unidentified mechanisms, and fragments of properties of different mechanisms[5]. Let's hold on, then, to the idea (in response to our questions about the role of modeling in today's science) that the modeling of phenomena can move in two di-

4 Taking this into consideration, we can see that the outputs of a given system have a double origin. One origin is the set of inner mechanisms within the system which generate outputs; but the factors, external to the system, which provoke the outputs, are another origin. And we also see that we may have recourse to two different types of explanation in order to explain the outputs of a system: we can explain these with reference to the factors which are external to the system and which provoke outputs, but also with reference to the mechanisms interior to the system which generate the outputs. Finally we now see that each type of explanation, taken separately, is not sufficient, because the outputs of a given system can vary with its inputs, and because the effects of a given input can vary with different systems. The notion of causal explanation is turned upside down by this point of view (see R. Franck 2003, pp. 184-186).

5 This is a burning question in the social sciences. I attempted to deal with it in R. Franck 2002, p. 93-98.

rections: the modeling of properties, and the modeling of the mechanisms for a given system.

6 Empirical models and formal models

We shall now see that the process of modeling mechanisms can also move in two directions. A mechanism has a material structure and a formal structure. The material structure is made up of a network of empirical relations obtaining between empirical subsets; in other words, these subsets and their relations are accessible through observation and/or experience. The solar system, as modeled by Kepler, is a famous example. But the solar system possesses another kind of structure, which was modeled by Newton. The law of gravitation, and the principles upon which it rests, do not represent an empirical structure. This law establishes a network of formal relations between concepts. It represents the formal structure of the solar system. And since Newton's model, as opposed to Kepler's, is not made up of empirical subsets and relations, but of formal subsets and relations, it can be applied to gravitational systems other than the solar system.

I began by saying that a scientific model represents a system, and that it represents the system's structure. We can now complete that statement. An empirical system has three kinds of structure. First, the empirical structure of the properties of the system. Second, the empirical structure of the mechanism which generates these properties. Third, the formal structure of this mechanism. The scientific model can represent any of these three structures of the system.

I suggest to call models 'formal' when they represent the formal structure of the mechanism of a system, and 'empirical' when they represent empirical structures, either of the mechanism or its properties. Strictly speaking, it is not the model itself which is formal or empirical, but the structure which it represents. For example, a model expressed in formal language is still empirical if it represents the empirical structure of the system[6].

6 The difference between empirical and formal models has been closely examined in R. Franck 2002, in the Conclusion of Part III.

7 Artificial models

Can we learn something about a natural object or process by analyzing the structure or the functioning of an artificial device built for the purpose of reproducing the behavior or the performance of that object or process? This question was asked by Massimo Negrotti last year in the introduction chapter to the first volume of the Yearbook of the Artificial. I will conclude this introductory chapter with a look at the question of the fruitfulness of artificial models in science. The method to which Negrotti refers has been used in research concerning artificial intelligence (AI). It has later been used in other domains, such as biology and economy. The devices can be virtual, that is, constructed by computer simulation methods. The computer model will be adjusted little by little to produce a better fit between its results and results generated by the natural system. When the fit is good enough, we suppose that the structure of the computer model is isomorphic with the structure of the natural system. And following this hypothesis, the model teaches us something about the natural system's structure.

The artificial device can be virtual, but it can also be material. The last is the kind of device for which Massimo Negrotti uses the term, 'technology of the artificial'.

> The interest for a general understanding of the artificial comes, therefore, from the persuasion that to *produce* something *ex novo* (let's name this *conventional technology*) is a quite different affair than *reproduce* something existing in the nature (let us name this *technology of the artificial*). [...] the technology of the artificial is devoted to replace natural *exemplars* with man-made substitutes able to exhibit some *essential performance* of them. (Negrotti, 2002, pp. 10-11).

Negrotti doubts that material devices can teach us anything about the natural exemplar whose performance they ape in some respect. Let's look closely at his argument.

An artificial device involves the use of materials which are different from those nature has used. The materials to be used in the construction of the artificial device enter the process with certain properties and qualities of their own. And unexpected internal interactions among the properties of the newly chosen materials can generate unforeseen side effects. Moreover, an artificial device cannot reproduce a natural exemplar in its entirety. Some observation level is selected in accordance with the particular performance or behavior of the natural exemplar which one wishes to reproduce. Negrotti concludes that while the core of an artificial device has to work in a manner similar to some particular natural thing, a number of further properties or performances, either actual or potential, may be activated suddenly by a variety of causes at several different observational levels. This is what Negrotti calls 'the unavoidable *transfiguration* of the *exemplar* and of its *essential performance*', brought about by the artificial. (id. p. 14)

The use of materials which are not the same as those which make up the natural exemplar, and the fact that the artificial device only reproduces a small part of the real system, leaving aside many possible interferences related to the multiple levels on which the real system may function – all this should convince us that the mechanism operating in the artificial device is often quite different from the mechanism of the natural system. And in this respect Negrotti is right when he says that the artificial device cannot teach us anything about the natural system.

Here are two examples which support his position. Armand de Callatay (2002, p. 108) writes:

> If an artificial heart is grafted to a patient, this heart is an engineering model of the natural heart. This heart uses different pumping principles. It does not use the energy molecules produced by the body. It does not self-repair nor adapt its size to the efforts required as a natural organ does, and it will not self-reproduce. It requires anti-coagulants in the blood. It performs anyway the same function.

In this example two materially different structures achieve a same performance. In a similar manner, we may note that it is unnecessary for a computer to resemble the human brain in order to mirror some of the properties of the brain. Thus in a general way we may state that similar performance may be achieved using different materials. And we may add that the same performance may be achieved through different mechanisms.

But an empirical system possesses three kinds of structures, as we noted. Apart from the structure of the properties of the system, the mechanisms of the system have both a material and a formal structure. A computer does not have to copy the material (neuronal) structures of the brain in order to mirror some of its properties, but doesn't the computer have to mirror the formal structure of neuronal mechanisms which governs the particular performances of the brain?

Under this hypothesis, the answer to Negrotti's question bounces back toward us. The artificial device does not teach us about the empirical structure of the natural system whose performance (in some respect) is reproduced by the device. But the artificial device can help us find out the formal structure of the natural system. It can help us do that if it is true that in order to produce identical performance, two different systems must possess the same formal structure, even though their mechanisms may operate with empirical structures which are different. In other words, two different empirical structures which produce identical performances should possess the same formal structure. This point of view reinforces our intuition, widely shared, that artificial devices, both virtual and material, are after all a good way to learn about natural systems[7].

If the point of view which I have just put forward is correct, the advantage provided by artificial devices, both virtual and material, is that they allow us to find out the formal structure of a natural system without knowing its empirical structure. They allow us to find out the formal structure of the system without having to open the box – to return to the metaphor of the black box. In what way? The model – more precisely the formal structure of the artificial device, whether virtual or material – can be adjusted little by little so that its results match those of the real system better and better, as I recalled just above. When the results match well, we can maintain – if the point of view I am supporting is correct – that the formal structure of the com-

7 If the artificial neural networks (ANN) are 'blind', as is sometimes said, because they are not constructed from knowledge of natural neuronal mechanisms, still they may be able to help us – if the hypothesis we are working under is correct – to uncover the formal structure of neuronal mechanisms. This is what I tried to show in R. Franck 2002, pp. 142-145.

puter model or of the material device is isomorphic with the formal structure of the natural system. Thus, it is the results, that is, the properties of the system, which guide us in the figuring out of its formal structure, with the help of an artificial device. Such a procedure does not seem different from some types of experimental control. And it justifies our going even further in the use of artificial models in scientific research, provided that we specify what knowledge we expect their use to yield (see Burch 2002, pp. 245-265).

Bibliography

C. Bernard, *Introduction à l'étude de la médecine expérimentale*, Garnier-Flammarion, Paris 1966. Original work published in 1865.

T.K. Burch, 'Computer modelling of theory, explanation for the 21st century', in R. Franck (ed.), *The Explanatory Power of Models. Bridging the gap between Empirical and Theoretical Research in the Social Sciences*, Kluwer Academic Publishers (Methodos Series, vol. 1), Boston, Dordrecht, London 2002.

A. de Callatay, 'Computer simulation methods to model macroeconomics', in R. Franck (ed.), *The Explanatory Power of Models. Bridging the gap between Empirical and Theoretical Research in the Social Sciences*, Kluwer Academic Publishers (Methodos Series, vol. 1), Boston, Dordrecht, London 2002.

R. Franck, 'Conclusions of Part I', in R. Franck (ed.), *The Explanatory Power of Models. Bridging the gap between Empirical and Theoretical Research in the Social Sciences*, Kluwer Academic Publishers (Methodos Series, vol. 1), Boston, Dordrecht, London 2002.

R. Franck, 'Causal analysis, systems analysis, and multilevel analysis: Philosophy and Epistemology', in D. Courgeau (ed.), *Methodology and Epistemology of Multilevel Analysis. Approaches from Different Social Sciences*, Kluwer Academic Publishers (Methodos Series, vol. 2), Boston, Dordrecht, London 2003.

M. Negrotti, 'On the logic of the artificial', in M. Negrotti (ed.), *Yearbook of the Artificial*, Vol. 1, Peter Lang, Bern, Berlin, Bruxelles, Frankfurt a.m., New York, Oxford, Wien 2002.

Robert Franck, Professor *emeritus*, Centre de Philosophie des Sciences, Université Catholique de Louvain

Massimo Negrotti

Models Thinking

Abstract:

The shared meaning of the word 'model' has changed a great deal during the last three centuries, shifting from something real to be reproduced to something artificial that tries to reproduce something real. This shift is probably due to the gradual discovery of the role of the mind in trying to reproduce natural objects or processes in different ways. In fact, today it is widely recognized that, when we describe the world, we are actually describing our models of it. This point has even been emphasized to maintain, as constructivists do, that reality does not exist at all.

More reasonably, it is possible to maintain that both reality and the human mind have their own rights. Therefore, when men build models for reproducing something natural, they have to take two complementary aspects into account. On the one hand, there is the capacity of our mind and our tools to simplify and, hopefully, to capture what is 'essential', at least at some observation level, in the real object under study. On the other hand, there is the certainty that what is left in the model is not a mere residual part of reality but, on the contrary, the largest part of it.

Keywords:
model, theory, methodology, artificial, technology, naturoids, science, nature

1 Models and theories

Models are attempts to reproduce some area of reality. Therefore, whatever stuff they are made of – logical, mathematical, statistical, symbolic, graphical, material etc. – they are true naturoids (Negrotti, 2002), i.e. artificial devices aimed to reproduce some aspects of the world. A theory is a propositional description of the world – regardless of the language adopted – which is unable, as such, to reproduce anything but only able to describe and to predict events in the world. In turn, models very often are built to generate events in the world – and, therefore, they claim to be intrinsically able to generate predictions thanks to the dynamics of the models themselves –, though they do not adopt the same 'materials' nature adopts.

Nevertheless, the meaning of these two terms clearly overlaps in some measure. Though the term theory, is again at the core of every scientific work, the term model, covers a large part of current discussion and work within every science, technology and even philosophy. In order to persuade ourselves we should have a look at the following fig. 1, where we collected the frequency of the two terms in question, within 23 disciplines, or, better to say, within Internet sites regarding these disciplines[1].

As we can see, the term 'model' goes well beyond the term 'theory' in areas regarding technologically based ways of working. In some of these areas, such as engineering, medicine, bioengineering or chemistry, it is easy to argue that the reference to theories lags behind the concept of model, giving models a characteristic of theory – derived tools. On the contrary, it is plausible to think that many other areas resort to models just because of a lack of established theories, like in the case of artificial intelligence or virtual reality. In any case, the role of the term 'model' is clearly gaining a wide relevance in all disciplines, except, as a trend, some classical natural and human sciences.

We could suppose that the concept of model is playing a growing heuristic role. Its adoption, as a way to reduce complexity or to introduce new concepts, tends to replace, at least as a research starting

1 The survey is only an introductive approach to the phenomenon we are dealing with. We simply asked at three search engines (in our case Google, Altavista and Yahoo, at the end of December, 2002) to count the Internet sites including the formula 'theory and physics' or 'model and physics', and so on. Of course, we cannot deduce, from such a survey, what role the terms in question played in the sites found. On the other hand, the table can be useful to at least make a general comparison among the selected disciplines. In any case, it has been found that in the Deutschen Bibliothek in Frankfurt the works adopting the term 'model' exhibited the following trend in the last decades of the past century: 1950: 30; 1960: 50; 1970: 350; 1980: 700; 1990: 1200; 1996: 1700 (Mueller, 2001).

point, the traditional theoretical discussion and related experimental work.

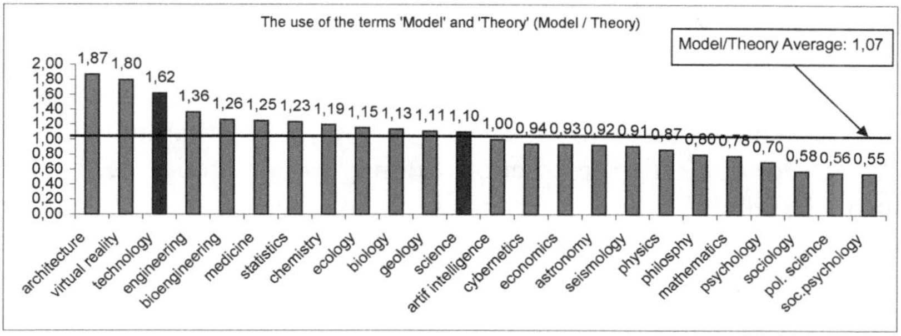

Fig. 1. Data were collected worldwide from Internet sources using Looknet© software designed at IMES-LCA, University of Urbino, Italy.

Resorting to a standard definition, available in the 'theory' entry of the *Oxford Companion to Philosophy*, a scientific theory is '[...] an attempt to bind together in a systematic fashion the knowledge that *one has* of some particular aspect of the world of experience' (*Oxford Companion to Philosophy,* 1995, italics added). Following two masters of methodology, Ernest Nagel and Carl G. Hempel, the above definition then refers to theories as 'hypothetic-deductive systems', in the sense that the available knowledge has to be deductively consistent with few axioms or hypotheses. Though we have a lot of different definitions for the concept of theory (Marradi, 1984), it is clear that, to construct a theory, we need to hypothesize something and to know something (empirical evidence or laws). Once the theory is constructed, it will allow us to make predictions which were not possible on the sole basis of the stand alone laws or hypotheses, and, therefore, a theory can always be falsified through 'crucial experiments'.

Models also need to be based on some established knowledge and some hypotheses, but it cannot be said that a model is a 'hypothetic-deductive system'. This is due to the fact that, particularly in the fields listed above, a model resorts to assumptions and conjectures rather than to high-level hypotheses on the state of the world as is the case for theories. More specifically, often models do not rely on generalizations on the matter in question, but on the description of *local* features of an object or a process. Even when a model is a theory-driven construct, it refers to a *specific* implementation of a theory and, therefore, it does not aim to give rise to new generalized knowledge.

Models are also conceived to make predictions possible – in the end, almost all of them are built for this purpose – but they do so without deducing anything from general laws, at least not in the sense that theories do. Rather, predictions coming from a model are the output of the interplay of the 'modules' (symbolic, logic, mathematical and even pieces of theories) they are made up of.

A good model is expected to work well, while a good theory is expected to be true. If a model works fine, i.e. it generates output fitting observational phenomena and their dynamics, then it may be assumed as a plan of action, rather than as a theory of the object or process in question. In fact, a model usually does not take the form of a formal proposition, which is necessary to have a theory. If it does so, then it should be called a theory and not a model.

Take an example from engineering. If we have to design a bridge over a river, then we build a model including all the aspects of the physical situation the bridge will have to function in: the river and its physical properties, the geological structure of the river banks, the expected traffic on the bridge, the most suitable materials to adopt, and so on. As a rule, if the project developed on the basis of the model is to work well, we will gain no new knowledge from it regarding the hydraulic laws governing rivers in general. Rather, we will gain, new expertise in making bridges. The hydraulic laws have to be known *before* we design the bridge. This is not always true, of course, because local cases always generate some problem – for instance of a hydraulic nature – and a new theory may emerge from the study of such a problem: but this is not the objective of a model for a bridge in itself.

Thus, from a methodological standpoint, models are not to be confused with theories. Rather, models can sometimes be a sort of prelude to theories, as it occurred in the famous case of Ernest Rutherford and of Niels Bohr's model of the atom, proposed in 1915, built through analogy with the solar system, which paved the way for the quantum theory. When models embed some theory, they try to establish order in a field. Therefore, they can be helpful for setting up new hypotheses and, therefore, to give birth to possible new theories.

The main uses of models, from a classical perspective, may be outlined in the following way, drawn from the brilliant, and even anticipatory sketch proposed by Ludwig Boltzmann in 1892 (Boltzmann, 1892). Models can be *instructional*, when they are concrete objects, perhaps built with gypsum or wood, metal etc., aiming to make the laws of nature visible; *calculating machines*, able to operate like humans in all the space of mathematical calculations; *models of theoretical physics*, conceived as mental constructions which only work in a physicist's mind until they receive some attention and induce actual realizations at some shared level[2].

2 Models as naturoids and their reliability

Due to the great explosion of the models approach in the second half of the XX century, even the lexicon adopted for speaking about them tends to be very different from the traditional lexicon of scientific methodology. For example, as we said, a theory or a hypothesis are expected to be true or false, while a model is expected to be reliable or unreliable.

Furthermore, while a theory has to be *tested*, a model should be *implemented* and then *compared* with the real object or process. This happens with the assumption that, if the model works well, then it can be adopted as a good approximation of the reality itself and, therefore, it can be studied *as if* it were the real object or process. The difference

2 Among the several typologies of models which have been proposed in modern times, it is important to cite the one by A. Rosenblueth and N. Wiener (1945), who simply distinguish *material* from *formal* models; the one by K.W. Deutsch (1966), who classifies models as tools for *organizing*, *heuristic*, *forecasting* and *measuring* aims and the one proposed by R.N. Giere (1999) who establish a rather complex typology, speaking of *instantial* or analogical models, *representational* models, *material* models, *abstract* models, *hypotheses*, and, finally, *theoretical* models. Apart from the obvious, partial overlapping of some definitions, it is meaningful to remark that, the complexity of the typologies grows over time, rather than converging towards some unified classification.

is an important one, because it is not so certain that a model could be conceived as an experimental tool or environment, while theories surely need to be tested through experiments.

In other words, if a model works well – i.e. it exhibits, through comparison, the features of the object or the process it claims to reproduce – no simple deduction can be drawn regarding its 'truth', though, from that moment, it can be considered as a reliable model. Since a model is always a complex structure of interrelated 'modules' – really it is a system and not a proposition – it may be able to reproduce what the 'theory of the artificial' (Negrotti, cit.) calls the *essential performance*[3] of the natural *exemplar* thanks to some special interplay or crosstalk between the component modules. This may have nothing to do with what happens in nature. In some ways, it is like a researcher who, using a model for inducing a known abnormal behaviour in animals kept in small cages by some stimuli could misunderstand the results of the study, because the abnormal behaviour could stem from the smallness of the cages rather than the stimuli protocol. Functional equivalents (Harnad, 1994) are at the root of these kinds of misunderstandings. In a word: clocks may be built in very different ways, and I would certainly be wrong if I tried to understand 18th clocks modelling them through the reverse engineering of a contemporary digital clock.

3 Models' complexity

Working on models is quite different than working on reality. Though we cannot work on reality without some model of it, reality responds with its own resources, quite ignoring our models. Indeed, we build

3 Every model pretends to capture something 'essential' which acts as an 'attractor' around which any other aspect becomes complementary. All the game is played, therefore, on the correctness, or on the plausibility at some observation level, of this claim. In general terms, 'The essence of a process [or of an object, NdR] is that without which it could not take place as it does; it is that through which a process "is what is"' (Frank, 2001). Or, let us add, 'what it looks to be' on the basis of a more or less shared judgement.

models in order to give meaning to reality and to gain further knowledge of it, but it seems that too much time is often spent defining, constructing, completing and revising models rather than observing real phenomena directly or indirectly and building or adapting new theories.

This is surely due to the availability of more and more powerful computers which allow us to implement even very complex mathematical, statistical, logical or symbolic models.

Here the methodological problems are different from those we discussed above. In fact, if we shift from a hard field of modelling (within which the implementation implies the adoption of real materials) to a soft one (within which the implementation implies the translation of hard phenomena in symbols) we suddenly feel a sort of liberation from the severe constraints of reality.

In some way, computer models are permanently correctable without high costs, and, therefore, they give the impression that reality is at our disposal: a sort of inexpensive virtual world within which we can reproduce, predict and modify anything, respecting and even surpassing natural laws.

But reality always awaits us at the exit door.

One of the common problems in simulation has to do with the boundaries of the model, i.e. the boundaries of the system we want to reproduce symbolically. It is clear that fixing the boundaries of a model of a natural system incorrectly, means losing all the features which, in nature, require phenomena that occur beyond the established boundaries. Of course, the concept of boundaries must not be considered only in topological terms, as when one tries to build a model of a river neglecting the structure of the river bed. The concept of boundaries implies the concept of *observation levels*: this means, for instance, that the *essential performance* of a natural object requires the contribution of disciplines x, y, z and w and not just x, y and z, if we want to build a reliable model.

There is much debate surrounding this aspect of our theme. For example, a specialist, writing about cancer risk according to a certain type of modelling, says that

> Such models involve parameters with biological meaning, such as cell proliferation rates, that could in principle be measured directly, rather than being estimated in a model-fitting process. It has been hoped that, by incorporating more biology, such models could provide more accurate estimates of low-dose risks (Crump, 2002).

Furthermore, boundaries may be conceived as special computational situations, usually at the extreme of input values or in the presence of strongly non linear equations. In these cases, only the experience (i.e. the expertise of the modeller as such and not of the scientist) can establish the 'domain of applicability' and the 'safe zones' of the model: the ones within which the output can be safely assumed to be reliable.

4 Use and abuse of models

Adopting models is unavoidable and useful not only for science and technology but also for survival. Humans as well as animals, generate models in various forms to map our territory, to anticipate what we are planning to do, to put ourselves in social situations and so on. In some ways, our life is permanently characterized by the building, refining and changing of models according to the feedback we get from reality as we perceive it. In any case, models are simplifications of reality, not only because of our limitations in perceiving reality but also because several classes of real phenomena are completely unknown, though they are likely to act on us in varying degrees.

Due to their simplification, models are always 'ideal' to some extent, since they tend to privilege, in every moment, what we think to be 'essential' in the target situation, object or process. There is a sort of 'eradication' of something, which we put in the foreground, from a whole which we leave in the background. It is a known fact that, in doing so, we are much more likely to make errors than to act properly. Of course, this is also true of scientific hypotheses and theories, but while models claim to reproduce the structure and the dynamics of natural objects, hypotheses and theories only aim to describe and explain them. Thus, it becomes clear that to describe and to explain the behaviour of the sun is quite different matter than to model the sun perhaps hoping to rebuild it concretely as occurs in the many fields of the current technology of naturoids, like bioengineering, robotics, etc.

On the other hand, the fact that models are simplifications not only of reality in itself but even of that part of reality we can perceive, means that there are clearly quantity and quality limitations in our capacity to rationally process the elements we would introduce in our models.

In turn, this implies that the most complex model of the World cannot be, as such, more complex than the most complex model of a little village. In other words, the maximum complexity we can assign to a model depends upon our limitations and not the intrinsic complexity of real things. If you assume the right observation level, then you will discover that there are not rich and poor areas of reality. Thus, the complexity of a 'simple' flower is quite beyond our modelling capacity just as the complexity of our brain is beyond our modelling ability.

Of course, scientific theories are also limited our own human limitations. But, when they are accepted as true, it means that the complexity of reality has allowed us, so to say, to grasp something actually happening inside it. As a consequence, the predictions based on established theories exhibit a reliability which stems from logical (deductive) reasons and not to something 'equivalent' to the causes which act within the phenomenon in question. In other words, if you want to predict a phenomenon you need at least a law governing that kind of phenomena. Equivalency or analogy – based on some isomorphism – in themselves are not some sort of alternate knowledge strategies but only, in some cases, a way to find out the right direction to take in our research into its specific nature.

On the contrary, as we have outlined above, there are several very different ways of getting some output, inspired by nature, from a device or a model. In any case, the price to pay is the impossibility of assuming all the different models to be true and generalizable reconstructions of the natural object or process, because this would be openly contradictory.

We can conclude that models are useful if and when they work as we expect them to work, regardless of their actual or internal consistency with the exemplar they are supposed to reproduce. Nevertheless, in order to guarantee that a model works well we cannot rely on the soundness of the model in itself – nor on the theories it possibly embeds – but on its final functioning as compared to what we expected. On the other hand, if a model is successful the clause of *ceteris paribus* will be much more menacing than in the case of a theory, precisely because we may not be aware which module, variable, relationship or coefficient, in the model, contributed decisively to produce the desired effect. This is a crucial point for all those inclined to think, in a rather mechanistic way, that modelling natural objects or process is only a question, albeit a complex question, of putting together the right pieces of a puzzle with the right proportion and the right relationships. Apart from several areas of technology, where the architecture of the puzzle is set up intentionally by man, there are no fixed rules to follow for building models whose knowledge content is reproducible in the same way a good theory has to be reproducible. Particularly in the social sciences, this involves clear responsibilities because, when a prediction, derived, for instance, from an interdisciplinary computer model, reaches the public, people do not discriminate the reliability of such a prediction from a prediction, let us say, made by an astronomer in his own field.

Bibliography

L. Boltzmann, *Über die Methoden der theoretischen Physik,* Populäre Schriften, J.A. Barth, Leipzig 1905, pp. 1-10.

K.S. Crump, 'The Linearized Multistage Model and the Future of Quantitative Risk Assessment', in *Human Experimental Toxicology*, 15, 1996, pp. 787-798.

K.W. Deutsch, *The Nerves of Government. Models of Political Communication and Control*, Free Press of Glencoe, New York 1966.

R. Franck (ed.), 'Introduction', in *The Explanatory Power of Models*, Kluwer Academic Publisher, Dordrecht 2002, p. 6.

R.N. Giere, 'Using Models to Represent Reality', in L. Magnani, N.J. Nersessian, P. Thagard (eds.), *Model-Based Reasoning in Sci-*

entific Discovery, Kluwer Academic / Plenum Publishers, New York 1999, pp. 41-57.

S. Harnad, 'Levels of Functional Equivalence in Reverse Bioengineering: The Darwinian Turing Test for Artificial Life', *Artificial Life*, 1(3), 1994.

A. Marradi, 'Teoria: una tipologia dei significati', in *Sociologia e ricerca sociale*, n. 13, 1984.

R Müller, http://www.muellerscience.com/NavStartenglish.htm, 2001.

M. Negrotti, *Naturoids. On the Nature of the Artificial*, WSPC, Singapore 2002.

The Oxford Companion to Philosophy, Oxford University Press, Oxford 1995.

A. Rosenblueth, N. Wiener, 'The Role of Models in Science', *Philosophy of Science*, 12, 1945, pp. 316-321.

Massimo Negrotti, LCA – Laboratory for the Culture of the Artificial, Università di Urbino

Danila Bertasio

Representation and self-representation: the concept of model in the arts

It's the painted man that eventually becomes the real man, the one alive. The other one, the original, the model turns to the state of remainder, of real burden.

A. Savinio

Abstract:

Art, which is just an incidence of truth setting itself into motion, produces models that are such in themselves. In other words, art is an expression of subjective representations that originate from the controversial relationships built around the artist's self and his inner life, with its sadly destroyed and happily integrated elements.

Thus, art portrays a way of living that is just a paradigm of itself. It gives us back reality by availing itself of the solitary strength of an individual's representation. In so doing, it often combines meanings and elements that lack an obvious connection, blending them into a complete and self-sufficient representation.

Keywords:
art, science, gallery, artistic representation, level of observation, artistic language, model, portraits, self-portraits

1 Nature as a model

The distinction between nature and art is not at all spontaneous and ubiquitous, but it rather reveals itself as a historically conditioned category. It is quite alien to the ancient world where there's no distinction between natural and artistic beauty; Aristotle's Poetica, for instance, is constructed without any specific reference to the idea of artistic beauty, which is only mentioned as a reflection of the first one. For the ancients beauty lies primarily in nature and when there's mentioning of it this is done with reference to that of the universe intended as Cosmos, a harmonious whole.

The impossible opposition of natural and artistic beauty is also testified by the theory of Memesi, even though not in its naïve versions – those telling about doves picking up Zeusi's painted grapes or that of the painted fly landing and trying to drive itself away as if it was a real one – but rather by the principle that establishes the total identification of the painted object with its natural model, in the sense of a copy or a replication.

The artistic act is an imitative one as it is a derivation of the natural pattern to which it is linked by a relationship of total dependence: all works of art are fascinating because such are their natural counterparts. In fact, it has to be stressed that it is not just the original Greek theory of the imitation as the following theoretical developments to emphasize on imitation as an act of natural derivation or direct approach of the living world, even though this searching has not to be pursued in the single ordinary individuals or events, but, instead, in all those models endorsing ideal perfection.

And in fact, among both the Humanists and the theorists of figurative arts the fundamental principle remains that imitation of nature as a direct belief that the human art is sort of a minor sister of the greater divine art. Niccolò Cusano, for instance, spoke about the human *ars finita* and its imitation of God's *infinita* one.

Nevertheless, the concept of imitation arises to a new meaning for it points out the importance of a direct and scrupulous approach to the natural objects, an aspect this which had been absent both in the ancient times and in the Middle Ages.

This is particularly evident in Leonardo, who compares the painter to a mirror and demands him to be universal master in counterfeiting all qualities traceable in nature-made shapes (Leonardo, 1995, nn. 53 and 73).

The idea that artistic beauty is a reflection of that created by nature is also realizable in Vasari who tells about the artist's obligation to nature (Vasari, 1981, vol. I, p. 369). Dürer on his hand, firmly stresses on its terms 'for the art really hides itself in nature and he who's able to draws it out of itself he will posses it' (Dürer, 1961, p. 53).

During the Renaissance period we do not really get to a real distinction between art and nature, an aspect extraordinarily witnessed by the collections of arts and wonders kept in the famous KunstKammer, those places where the history of an object was reconstructed from its natural forms to its artificial representations and extensions, even those in which it was aesthetically transfigured.

These wonders' chambers were actually really fantastic places devoted to the *machinamenta artis et naturae* in which men of great intelligence created order and beauty out of chaos and turmoil and in which it was possible to understand the principles on which the world is set. They reflected the spirit of the times which was sensitive to Encyclopaedism, legitimate heir of that Middle Ages' Scholasticism whose aim was, on the basis of its strong appeal on all major European royal families, to reassemble the universe and grasp the unity of its structure[1].

All this dramatically ended in 1720 with the dismemberment of Dresden KunstKammer, an episode that drew the attention on the latest embellishments of a slow process that radically modified the relationship man had established with the world around him. Their decline coincided with the birth of a new science and consequent renouncing to the old dispositions to look for confirmations of one's own views about the world through human senses. Minor qualities, those directly connected to the latter ones (colours and tastes) lost ground in favour of major ones, suitable of being mathematically investigated (mass and weight). Science was now willing to deal with what could be objectively studied, thus excluding all that had to do with subjectivity. The world is written in mathematical characters even though numbers no longer have that sacred and harmonic value, typical of the Pythagorean-Platonic tradition, to assume the guise of calculability.

1 It is not just a case that only those great families made up of kings, queens and emperors could afford these precious and very expansive 'pieces of world' that were supposed to reveal its mystery. Among others it will be good to mention the collections by the Valois, John of Berry's treasures, those of Charles I of England, Leopold Wilhelm and Rudolf II.

The unity of knowledge wanes definitively and the art is not or rather is no longer 'truth setting itself in motion' (Heiddeger).

From a cultural point of view, two antithetic positions arise: one which focuses on the compensative role of the aesthetic experience, the other stressing on the implicit impoverishment deriving from the dualistic opposition between a scientific knowledge of the world and an aesthetically perception of it – arisen to a new level in the contemporary world.

The theory of ideal beauty will laboriously survive from the ancient times through the eighteenth century and will experience a last important revival with the age of Neoclassicism. By introducing an element of artificiality when producing the copied object, beautiful nature, so much speculated about by all modern theoreticians, becomes something different from the observable nature, partly because it represents an archetype that lies beyond things and partly because, in its extreme versions, it will end up coinciding with art itself.

The Romantic movement will end up being the last and losing attempt to oppose the model of reality offered by the sciences. And it's not just a case that in the arts we witness, during this period, an intentional abandoning of the imitating principle by most artists. Karl Phillips Morritz wrote in one of his essays dated 1788 that art is beautiful not just because it is a reproduction of nature but most of all because it is a creative force originating from it. Therefore, the imitative action doesn't belong to the *naturata* nature but, instead, to the *naturans* one. Wilhelm Schlegel will eventually counterfeit the traditional theory of imitation right on the basis of Morritz's idea which considered the artistic force as an independent and creative one. This position is also shared by Schelling to whom art, being comparable to nature for its originating value, brings the artist to disregard the nature-made products in favour of the creative ability concealed behind them: only this way he will give birth to something alive.

The crisis that followed opened to way to the mirror-like conviction, very popular during the nineteenth and twentieth centuries, that art becomes the model of itself, the source of all beauty, artificial and natural alike. Typical of this approach is the example of Monet who while aiming at painting what he saw ended up seeing what he was actually painting. Just think of his private garden at Giverny which had been built on the model of his masterpieces.

2 The theme of 'truth' in the arts and sciences

The art, which isn't just a case of truth setting itself in motion, produces models that are such in themselves or, to say it differently, is an expression of subjective representations originating from the controversial relationships built around the artist's Self and the sadly destroyed or happily integrated parts of his inner life. Thus the art portrays a way of living this world which is just a paradigm of itself. It gives us back the reality we live in by using the solitary strength of an individual's representation. By doing this it often combines meanings and elements apparently lacking in evident connection and it blends them into a complete and self-sufficient representation.

All this does not imply that it has to be meant as an irrational form of knowledge, but rather as a differently oriented way of rationality. The art, just like science, considers its confidence in free interpretation as the centre of its action, but it is able to avoid at the same time the validating or invalidating power deriving from the observation of reality to which science is irreparably bound.

Due to the differences here mentioned, which are concisely the basis of both the artistic and scientific philosophy, it is common opinion that for the western social society of today the configuration assumed by the art and science is distinguished by a certain impoverishment of knowledge, a sad renunciation under a more generic cultural profile, but most of all it is source of degrading for the art. In the comparison with the scientific thinking the artistic one appears to be arbitrary not only for the pervasive election of pragmatically oriented criteria during the practice of our faculty of wisdom, but also and in particular for it is the source of outcomes which are mostly not comparable between them.

In short, on one hand we find the exploratory nature of the scientific method, that considers induction as an essential moment of the investigation but also acknowledges the ineliminable role undertaken by deduction in the exposition of facts, and which brings man to give up to the idea of his being at the centre of a universe that no longer guarantees certainties but instead a variety of problems of interpretation.

On the other hand we find art that, no more subjected to the criterion of truth, doesn't have the necessary power to avoid that such a renewed idea of man falsifies his own centrality.

In conclusion, while it is reality with its own empirically based rules to impose itself to the man of science, in the artistic domain man still insists in imposing an interpretation based on its own interior models of reality.

As a matter of fact, both the scientist and the artist watch the world carefully, both equally seeking a balance between imagination and custom, rule and chance, reason and instinct, but the result they both achieve cannot be superimposed being theory what comes out of the former ones and art what emerges from the latter ones.

Even though what they both aim to is intrinsically different, it is not possible to detect substantial differences between the two as far as the cognitive process is concerned for in both cases the investigation and consideration over reality are basically shared. The scientist, just like the artist, privileges his preferences, motivations and personal intuitions. For both the achievements of their speculations, devised in the lonely spot of his lab or that of his *atélier*, have a similar epistemological meaning which can be traced just in their common effort to elaborate a new vision of the world or new rules to see it.

And yet, even though in their own individual and private context they share this same attitude and similar thirst for knowledge, the result of their efforts is inevitably different.

The scientist will consider reality as his supreme judge: this will be sharp and cruel, totally indifferent to his belief, hopes and hard work. It will cut shortly his theory by simply state it as true or false.

The artist and its work instead will be judged by another tribunal, that of his hypothetical audience that might not share his vision of the world or most likely define the outcome of his artistic effort as beautiful or ugly, understand it or blame it, display it under the lighted vaults of a museum or simply forget about it. But no word about truth or falseness will be spoken and the artist's subjectivity will survive and avoid any sort of cage. He will nourish his hopes as he awaits for new generations willing to appreciate his way of watching the world. His work will remain to give birth, through other works, to more powerful and more investigating rules of his own way of perceiving reality until in the end the audience as well will acknowledge that it's possible to get to know the world that way too.

The artist might not achieve fame and fortune, but nobody will ever deny him the possibility to imagine future sharings. His work of art will stand the test of time, it won't suffer the evolutions of taste and fashion, but it will stay there awaiting even for that only single person in the world able to appreciate his own unique expressive style (value).

3 The artist's gaze and the model

An *oeuvre d'art* is a subjective form of knowledge that just like any scientific theory is inclined to be compared to reality, even though on the basis of a different opposition, not true/false but beautiful/ugly. And in fact, its freedom from the typically eager scientific criteria of truth is proudly confirmed by the artistic unique ability to combine together multiple varieties of expression.

In this respect portraiture further emphasizes the artist's interpretative free will, for beyond the mere configuration of an optical phenomenon artistically reproduced, he aspires to unveil what lies underneath the surface.

In other words, the goal the painter tries to fulfil is that of a work in which the features' unity of a face is strongly reinforced and closely examined not just in terms of merely empiric phenomenon – his physical representation – but most of all by operating a harmonic junction of those expressions and movements that, in his opinion, specifically allow the subject's inner soul to shine through, to reveal the structure of the body and its dynamism, to clarify the relationship that man has most likely established with the world around him (Simmel, 1985, pp. 51-62).

If nowadays' aspiration to be portrayed is regarded as a sign of egocentrism, the act of doing it for real is certainly less understandable as such a choice would publicly be considered index of a snobbish and old-fashioned mentality.

And in fact, mr. Anybody's decision to be portrayed, as did Giovanni Agostino della Torre (1515, National Gallery, London), would cause more than a smile in contemporary society.

His decision would be regarded as a bit over the edge even in the case he claimed for himself, just like did Lorenzo Lotto, a posture signifying solid professional authority, one totally free from self-adulation, and as only reference to his job a cartridge identifying him as Esculapius, the Greek god of medecine and recovery in his right hand and a copy of Galienus – physician and anatomist whose high reputation remained intact throughout the Renaissance – in his left one.

Fig. 1. Lorenzo Lotto, *Agostino della Torre,* National Gallery, London

It is right to underline that seeing today a nicely silver-framed picture, showing his/her family, in the office of some professional man certainly conveys (the female disposition is probably more inclined to this) a sense of reliability and calmness to his persona.

We would certainly perceive a feeling of somebody extremely vain if the same mr. Anybody was caught in a picture portraying him on the wind and water line of some sunny beach or, even more, if we saw his portraiture hanging on the wall behind him, maybe one clearly showing a compendium technique[2].

If today things have changed and a hunger for visibility is lost in the one-thousand different brooks made available by the so-called 'advanced society', nevertheless it has to be recognised that among all human vanities none has ever been more captivating for the human imagination than the idea of possessing his/her own portrait and to make of it a model of human being destined, more than any other possible object, to stand the test of time and enjoy eternity. And as a matter of fact, today still, to have got a portrait of ourselves gives us the illusion to be able to escape oblivion and to overcome the silence of death. In this respect, it will be enough to say that we know Giulio Cesare's or Dante's faces for their portraits have accompanied our history. We even went as far as to give God a face, something that ultimately led the Church to win the battle against Iconoclasm in the name of a presumed likelihood.

'A man's whole life can be a lie for himself and the others [...] nevertheless, a picture of him painted by a great artist can translate the real character onto the canvas and reveal his secret to all generations to come', wrote William Hazlitt at the end of the age of the Enlightenment.

And this was even more evident starting from the Renaissance period, in that particular historical moment when Leonardo fought for painting to be considered as worth as poetry and for it could be re-

2 The compendium technique, – writes Daniel McNeill – together with the model's vanity and his ambition to refer to the posterity, tends to convey to portraits a sense of quiet formality. The model stares at nowhere, mute and immobile, it's simply a face turning into eternity. This stillness partly derived from long hours of posing and if it helped the work of the artist on one hand, it was also responsible for the lost of expression on the other (McNeill, 2000, p. 147).

garded as something capable to represent any action in a way as exhaustive as only poetry could do.

Painting, according to Leonardo, was able to truly express its moral aim by depicting the human action by means of gesture and facial expression, and that's what Leonardo says about it:

> may a poet say I create a fiction narrating great things, he himself will be a painter just like Apelle did slander [...] if poetry becomes moral philosophy, then this [painting] is natural philosophy for while the former describes the *operationi* of the mind, the latter focuses on how the mind is reflected by the movements [of the body]: if the first one frightens people by means of *infernali fitioni*, the other one does the same by visualizing them (Blunt, 1966, p. 65).

The consignor commits his vanity certainties and uncertainties to the painter's hand, with the same sort of trust he grants his hairdresser. And he'll do it well into the XXI century where this genre of artistic production is constantly required. It will be enough to remember that the lack of this artistic form from museums had been highly criticized by the artists themselves.

Hanskell underlines that until 1888 there had been not more than six portraits out of two-hundred and eighty paintings held at the Musée de Luxembourg. That means about 2% of the grand total and clearly reveals how little that museum represented the painting of the age, particularly when you think that at the Salon exhibition of that same year five-hundred and seventy-seven portraits out of two-thousand and eighty-six paintings had been displayed, that means about 22% of all, ten times more than you could actually see at the Musée des Artistes Vivants (Kanshell, 1987, pp. 221-222).

Quite obviously the kind of portrait released by an artist isn't the same one you could actually expect from a candid camera, but rather the result of a combination of both technical ability and a particular quiet complicity developing between the artist and the model, both equally engaged in the undertaking of what Goffman describes as 'self-presentation'. For this reason too the specific artistic genre deserves a particular attention. We can imagine for instance that Mr. and Mrs. Andrews were certainly satisfied at seeing their image, so well mannered and coherent with the manuals of the well-to-do of the time, portrayed by W. Gainsborough (see Fig. 2).

And we can certainly imagine how pleased Mr. Andrews must have been to see the quiet expression of his face and the representation of his well run lands, which would soon turn into an extraordinary visiting-card to the eyes of those friends coming to visit them at their place. His wife as well, we can guess, could not but appreciate her tidy, orderly figure and the silver-blue dress artfully revealing her pointed silky shoes.

Fig. 2. Thomas Gainsborough, *Mr. and Mrs. Andrews*, National Gallery, London

The same could be said about cardinal Bembo (National Gallery, Washington D.C.; see Fig. 3) as portrayed by Tiziano.

The cardinal emphasizes and amplifies his speech by the calm gesture of his open hand, the palm upwards, his thumb straight and all other fingers bent. His hands, not less than his face, are at the centre of the representation, fully illuminated, thus becoming an essential instrument of communication with men other than with God. And again, actress Sarah Siddons could not but admire the image of her painted by Joshua Reynolds (London, Dulwich College), at the time the highest paid painter of the United Kingdom (see Fig. 4), here disguised as the fascinating muse of Tragedy, and find onto it all her dreams of success.

Fig. 3. Tiziano, *Cardinal Bembo*, National Gallery of Art, Washington

For this peculiar kind of complicity, today we still wonder why the Spanish royal family and most of all the queen mother, didn't order to destroy the work of Goya, highly esteemed painter of Court; instead they allowed it to be displayed in one of the main splendid halls of the palace.

Goya in his portrait 'Charles V and his family' (1799; see Fig. 5) had painted all the ills of the world onto the faces of Charles' children thus unveiling them to the eye of those at Court as the result of unfortunate and repeated marriages among kinsmen, a quite usual habit among the European families of the time. Their frocks, excessively gorgeous, emphasize their dismayed, idiotic faces. The artist not only was not dismissed from his prestigious office, but, instead, gained so much popularity that over four-hundred noblemen and members from the royal family asked to pose for him, all of them oddly convinced that his incisive style would make them immortal.

Fig. 4. Joshua Reynolds, *Sarah Siddons*, Dulwich College, London

Maybe Lord John and Lord Bernard Stuard, cousins of Charles I of England, were deeply surprised but at the same time happy with their common portrait by Antoine Van Dyck (1638 circa, National Gallery, London; see Fig. 6). The two brothers are here portrayed as if they are about to leave, maybe while waiting for their carriage to arrive in a moment. Lord Bernard, the youngest of the two, is depicted with his foot on the step, his gloved hand on his hip, glancing at the seer. Lord John on the other hand is elegantly leaning against a column, his expression almost revealing that we're in front of a contemplative soul. For convincing as they are on the canvas – writes Erika Langmuir – these characterizations of models could be deceptive. That Lord John here displaying a contemplative air, was described after his death as 'the most hot-tempered and harshest personality of those belonging to that illustrious family of princes' and that 'disliking the lap of luxury, he had chosen the military life'; on the opposite side, Lord Bernard, depicted in this portrait as a haughty man, was reported to be 'extremely gentle and kind by nature' (Langmuir, 1997, p. 198).

Fig. 5. Francisco Goya, *La famiglia di Carlo IV*, Museo del Prado, Madrid

Fig. 6. Antonie Van Dyck, *Lord John e Lord Bernard Stuart*, National Gallery, London

4 What with portraits and self-portraits

Things get a little more complicated in the case of self-portraiture. If in any portrait the model can recognize his own individuality at any moment and re-trace the aesthetic legitimated interpretation that has been given to his own physical appearance, in the case of the self-portrait the artist exposes himself to a very particular sort of traps and enigmas which are naturally implicit with the action of transferring outside something exclusive and subjective.

We could easily trace down the first examples back to the fifteenth century. The *Coronation of the Virgin* by Filippo Lippi (see Fig. 7), an altar-piece he worked on between 1441 and 1447, is sort of premise to the future maturity of this pictorial genre.

Fig. 7. Filippo Lippi, *Incoronazione della Vergine*, Galleria degli Uffizi, Firenze

It is sufficient to observe the image of Filippo which, although introduced to the seer together with other characters, is ideally positioned inside an enclosed space built to host his only presence. Lippi paints himself among a multitude of angels, saints and devotees, but his is the only glance turned towards the spectators, his right arm up and his fist supporting his cheek. The attitude chosen to represent himself contributes to isolate him from the other subjects represented and suggests, to those watching, the artist's effort to offer specific elements marking a difference between himself and the other characters portrayed, with whom he's sharing an intense devotional moment.

If Lippi isolates his own figure thanks to his devotional attitude, Lorenzo Lotto in his 'Cristo vite e il martirio di santa Barbara' (see Fig. 8), commissioned to him by the counts Suardi of Trescore (Bergamo, Italy) in 1524, places himself apart from the rest of the Suardi family kneeling in front of the figure of Christ, but at the same time he underlines his physical presence by using his signature.

Fig. 8. Lorenzo Lotto, *Cristo Vitae e il martirio di S. Barbara*, Cappella Suardi Trescore, Bergamo

The self-portrait as a work of art of its own – writes Stoichita –, on rectangular or circular panel forms, on board or canvas, displaying the author's half-torso, makes its appearance only at the dawning of the modern age. The self-projection within a fictitious context, particularly in a *historia*, is, on the contrary, rather usual since the age of the Greek classical art [...]. Before the XVII century, a period in which everything was re-discussed, the self-projection of the painter is almost uniquely contextual (Stoichita, 1998, pp. 202-203)[3].

3 For what concerns the theme of contextual self-projection in painting Stoichita offers a four-type model of possible representations:

1 *the contextualized author* typical of the middle-age miniatures. The pictor/scriptor is represented within the space of the letter he's writing/ painting. An example among others cited by Stoichita himself is taken from a 'Leggendario' from the XII century, cod. Bodmer, 127, which is held at the Bodmeriana Library in Geneve and that was executed by Brother Rufilus.

2 *the masked author*, a form of painting very popular during the last part of the Middle Ages and the early Renaissance, in which the painter is also the actor of a *historia*. An example can be found in the work by Benozzo Bozzoli *'Corteo dei Re Magi'* held at Palazzo Medici-Riccardi or also in the *oeuvre* by Lippi already cited elsewhere.

3 *the self-portrait with mask:* if in the previous form of presentation the autobiographical nature of the work is guessable from the glance of the subject which, in this case, is turned towards the seer while the other members inside the *historia* are totally indifferent to his presence, this other expressive device allows us to detect the author only hypothetically. The painter usually emphasizes his presence within the painting by adding his name right above his image. This choice of expression will last well until the XVII century. According to the historian the extreme example of this possibility is provided by the

As the historian remarks, it will be necessary to wait another century before the artist comes to dedicate his own work all to himself.

The self-projection of the author arises right from the very beginning several problems starting from its own premises: it is the one case where only the artist who has executed the work can recognize if he has or not succeeded in transferring into the painting that subtle and fragile relationship existing between the inner representation he has of himself and the image he sees in the mirror. In other words, he is, in spite of his technical ability, the indisputable judge, the only seer, model's model, who can re-trace himself in the representation he has given of his *persona.*

This is a peculiar aspect that self-portraiture shares with autobiography and that is, for example, clearly explicated by Rousseau while he was just about to start writing his *Confessions*.

He pointed out, in fact, in the notes of the Introduction that his aim was to offer the reader the story of 'a man in all the truth of his nature; and that man will be me. Me only. I can feel my heart and I know men. I am not just like any of those that I have met; I dare believing I am not just like any of those living'.

The self-portrait stands as a description, a praise about the uniqueness of a man who's about to become model in the world and of the world itself. The auto-biography does not tell anything, but describes an intimate and immediate state of the narrator; in fact, only afterwards, this can be meant as 'a story' of his personality. Only a chronologically ordered *ensemble* of self-portraits could have the same autobiographical meaning. In this respect the self-portraits painted by Dürer and those by Rembrandt as well stand as enlightening exam-

---

'Davide con la testa di Golia' by Caravaggio, now located at the Museum of Villa Borghese, in Rome.

4 *the self-portrait as a visitor* sees the painter introducing himself as a stranger element of the story he narrates. An example is that given by Dürer in *'Il martirio dei diecimila cristiani'* dated 1508. At the centre of the scene yet in a space free from other characters, we can see two figure dressed according to the fashion of the time. They are the author himself and one of his friends, the poet Conrad Celtis. In order to underline his presence the artist holds a pennant with his signature: *Iste faciebat anno Domini 1508 / Alberto Dürer alemannus.*

ples. The first of the two, starting from the one he made in 1484 in which, as he also will eventually write on the left margin of the sheet, chose to represent himself as a young men. This will be followed by 'Self-portrait with fur' that bears the writing: *Albertus Durerus Noricus / ipsum me propriis si effim / gebam coloribus aetatis anno XXVIII* (1500; see Fig. 9) and, at the end of his days, by that 'Self-portrait naked' (1522; see Fig. 10) and by a drawing titled 'Man of sorrow' (see Fig. 11) in which Panofsky will realize to be in front of a painting collecting the artist's presages of death, here retraceable in the superimposition of the two representations of Christ and of the mourning man.

Fig. 9. Albrecht Dürer, *Autoritratto con pelliccia*, Alte Pinakothek, Monaco

Rembrandt, on his hand, tells us about himself with a look full of wonder in his self-portrait dated 1628, dressed up to the nines in that painted in 1640, looking like a king in the one produced in 1658 and terribly sorry in the final one of 1665, where he depicts himself as an old artist 'no more able of joy or sorrow, no more willing to look at the world with that curiosity which allows things to emerge in their uniqueness [...]. He's staring at himself but his look is lost in nothing' (cf., Rella, p. 63).

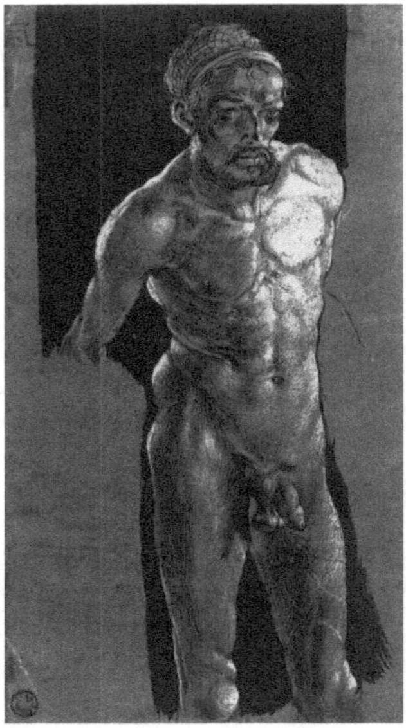

Fig. 10. Albrecht Dürer, *Autoritratto nudo*, Schlossmuseum,Weimar

Fig. 11. Albrecht Dürer, *L'uomo del dolore*

No model can actually make the undertaking easy and for this reason, more dramatically here than it is in portraiture, the artist produces this visual testament with the awareness that his identity is in danger.

The mirror is the only chance for him to grasp the image of his fragmented identity; it is the only object capable of helping him to re-collect his elusiveness. A banal mirror placed between himself and his own reflection, that makes it possible for him to recognize himself for

> nothing is more difficult to imagine than our face when there's no mirror to return the image. If I try to look at myself I find nothing but fog, night and abyss, as if the First person, taken to the limits, turned to a neutral anonymity or exploded in a multitude of empty masks. (Marquet, 1996, IX-X).

It is in fact the mirror that helps him to replace the question 'who are you?' with that 'who am I?'.

We're the only experts about ourselves and nobody else knows our thoughts, our aspirations or story as well as we do.

For this very reason a self-portrait is no portrait; it is the annihilation of every memory about the rules of representation for to represent our own face it is not the same thing as representing that of the one who's in front of us, male or female. In the case of portraiture all roles are very well defined with no chance for superimposition: the model is the only subject of the painting that the artist is creating. The representative aim is regulated by an interest and attention to the outward appearance, to the psychosomatic features of the face and body of the model.

In the case of self-portraiture instead, we face the affirmation 'this is my face' replacing the 'this is your face' one. Within this adjectival substitution we can trace a fundamental difference between two painting genres, a difference which does not rely on the fact that both focus on the outlook, on the outward features of the face or the curves of the body (cf., Boatto, 1998, pp. 3-7).

From an ideally point of view a self-portrait lacks in purchasers as these are the artist's Self. The very same moment when he decides to isolate his own portrait from the rest of the context to represent it in an *oeuvre d'art* implies the existence of a far superior level of self-conscience.

Besides, this self-reference is encouraged by an economical independence that the artist has started to achieve right during those years and from the consequent acknowledging of his new social role.

A striking example of this can be found in the painting by Pieter Paul Rubens (1606, Alte Pinakothek, Munich), 'Self-portrait with my wife Isabella Brant', where there's no hint to his job – no brushes, no canvases or easels. His luxurious dresses and a certain aristocratic allure are sufficient to communicate his economical success and social prestige to the seer. No memory, nor mentioning to the typical artistic sorrow which had always been a *topòs* of several biographies.

'Nobody's commissioned this work to me, I'm doing it for myself'. A *self in itself* or *for itself* is the exclusive aim of the self-portrait. This explain its being a form of self-referring expression, that cannot come to a simple and mere transcription of what the artist sees reflected in the mirror even because, already from his planning, he is fully aware he's going to leave an inheritance destined to cross the barriers of space and time, far beyond any possible cultural difference. The character in itself is already within the frame, the subject of his subject: it is the final outcome at the end of a long path populated with schemes and corrections and that for this same reason, cannot be conceived as a carefully process of visual recording or expressive experience. Instead, it has to be intended as an attempt to build a rational model which yet remains not duplicable for it involves just the artist producing it.

The final goal of the self-portrait is thus to investigate in the depths of the artist's inner soul, deep down into the most intimate and hidden folds of his persona. The artist aims at defining his own identity through the conjunction of his outward appearance with the invisible side of his 'Self', and in doing so he certainly takes his own risks for in the end, despite the mirror and his personal technical ability, it is possible for the 'Self' depicted to appear as even more inexplicable or strange to his own eye.

Let's just think, for example, what sort of influence can have on us the fact of being watched: the other sees us, reads us, foresees or anticipates our possible reactions and answers, he reflects about our movements, gestures and attitudes: he or she can get to know us or simply recognize us. We do have the possibility to do all this while we interact with our fellows, but not with ourselves. We can certainly watch many features of our body, our arms, our legs, our clothes, our figure from head to toe, but we cannot see our face and the way it changes in the blink of an eye or even how it can be marked by sorrow.

We're all blinded when it comes to relate to someone else in front of us. If we could be granted the ability to look at ourselves and regulate our behaviour on the basis of a permanent surveying over our visual communication and facial expression, our actions would lose in simplicity, besides without even any guarantee that our self-control can be estimated by the interlocutor as the best possible.

No man can tell who he is, said Camus, and his same sentence had been proved right four centuries before by Rembrandt in his painting on board 'The painter in his studio', dated 1628 (see Fig. 1), now held at the Boston Museum of Fine Arts. In it the artist choices to arrange his self-presentation inside an empty studio. At its centre stands the easel supporting a canvas which is destined to remain unknown to the seer for it can be seen from its backside only.

The artist is located at a certain distance from them, wearing a working outfit, way too large for his tiny size, and his face is shadowed by a hat. In his hands he holds his brushes, the palette and a handrest. Nothing, other than the canvas supported by the easel and the paint itself, can catch our attention, and, it has to be stressed, not as elements in themselves, but for the distance separating them.

The artist stands right in front of the spectator while enjoying a moment of thoughtfulness, lived at distant range as if to allow himself to rationally consider the image of himself he has produced. He looks at the painting while, at the same time, he glances at his face reflected in the mirror.

The seer, on his side, sees the artist but not the image of himself he's producing: 'It is the *"in machina"* gaze so typical of any self-portrait. The device Rembrandt is looking at is the instrument returning his own image: we are occupying the mirror's place' (Stoichita, 1998, p. 238).

A century before Parmigianino too proved to be also aware of the human limits to self-knowledge. His, in fact, is no self-portrait, but rather, the representation of a man caught in the act of self-investigation. He doesn't aspire to come to conclusions nor he tries to impose a model of himself, but instead, leaves this task to the seer. Unlike Rembrandt, Parmigianino allows the consciousness about the ambiguous interlacing between the artist, the mirror and the paint come through, emphasizing the ambiguity without resolution, which is implicit with self-reference, and ends up preferring self-reflection. This is the artist's way to underline the fact that his gaze won't be able to perceive anything more than what any man can see. For this very reason, he decides to turn his work into a mirror and himself in anything more than a man looking at his own image reflected in the mirror.

This *oeuvre* is no doubt one of the best examples, inherited from the XVI century, witnessing, possibly for the first time, the presence of the mirror, that subtle link interweaved with tricks taking its place between the artist and his work.

Vasari wrote right about this work:

> Further than this, in order to investigate the quibbles of arts, he started one day to paint himself by looking at his image as it was reflected into a looking glass, like one of those, half-circled, you can find in a barber shop: whilst doing so, his attention was caught by the distorting effect that the roundness of the reflecting surface was playing on every thing that made up the scene, and was thus encouraged to counterfeit every aspect according to his whim. Nevertheless its level of likelihood, including his own image, was so accurate and neat you could not estimate or believe. (Vasari)[4].

Vasari explains this executive choice by Parmigianino as a clear intention to produce the illusion tha the painting is nothing more but a mirror[5] – and also realizes that this mirror/portrait becomes, for the artist, an extraordinary chance to play with himself.

The looking glass reflecting the elegant features of the painter underlines, both for the artist and the seer, that the outcome could be

4 The statement by Vasari according to whom the kind of mirror painted reminds of those used by barbers is important for it testifies that the Flemish tradition was still unknown.

5 It must not be forgotten that that by Parmigianino was indeed a true discovery for the Flemish device of the rounded mirror was still ignored by the Italian tradition of his times.

nothing but the representation of a man looking at himself in a mirror, and that is due to the ineliminable problem associated with the relationship reality-illusion which, for the facial axis modifying the natural symmetric arrangement of the figure, remains inevitable.

The mirror, in facts, reflects a symmetry which is re-arranged but also turned up from right to left compared to the linearity of the model. The artist lies right in front of it fully aware that not even the mirror will allow him to see himself as all others do.

This limit is considered unsurpassable by the Flemish Van Steenwinkel as well. His self-portrait, now exhibited at the Antwerp Museum des Beaux-Arts, though not just a mirror, unlike that produced by Parmigianino, since it also reproduces elements of vanitas, cannot be described as simple still life, but rather as another example of cultivated artist caught in the act of gazing at his image in the mirror.

The model of Self reflecting in the mirrors forces the artist, at the very same time when he illusorily imprisons his appearance, to also surrender to the impossibility of its capturing, aware that this choice is, besides, the only possible compromise between the idea of Self and its self-acknowledgement, removing at the same time any analogy, likeness or resemblance with the rest of humanity.

Bibliography

A. Boatto, *L'autoritratto moderno da Goya a Warhol*, Laterza, Bari 1998.

A. Blunt, *Le teorie artistiche in Italia*, Einaudi, Torino 1966.

L. da Vinci, *Trattato della pittura*, TEA, Milano 1995.

A. Dürer, 'Proportionslehre', in Id., *Schriften, Tagebücher und Briefe*, Stuttgart 1961.

P. Fossati, *Autoritratti, specchi, palestre*, Mondadori, Milano 1998.

F. Haskell, *Le metamorfosi del gusto*, Bollati Boringhieri, Torino 1987.

E. Langmuir, *National Gallery*, National Gallery Publications Limited, London 1997.

C. Le Brun, *Le figure delle passioni*, Cortina Editore, Milano 1992.

D. McNeill, *La faccia*, Mondadori, Milano 2000.
J.-F. Marquet, *Miroirs de l'identité. La littérature hantée par la philosophie*, Hermann, Paris 1966.
F. Rella, *Negli occhi di Vincent*, Feltrinelli, Milano 1998.
G. Simmel, *Il volto e il ritratto*, il Mulino, Bologna 1985.
V.I. Stoichita, *L'invenzione del quadro*, il Saggiatore, Milano 1998.
G. Vasari, *Le vite de' più eccellenti architetti, pittori et scultori italiani, da Cimabue insino a' tempi nostri*, Einaudi, Torino 1981.

Danila Bertasio, LCA – Laboratory for the Culture of the Artificial, University of Urbino

Giovanni Boniolo

Theories and models: really old hat?

Abstract:

In this paper the topic of the relations between scientific theories and scientific models is tackled by considering the former as hypothetical scientific representations *and the latter as* fictive scientific representations. *A classification of the models is also proposed.*

Keywords:
theories, models, representations, fictions

1 Introduction

The issue of the relations between scientific theories and scientific models might be considered as old hat; something on which nothing new can be said. However each different interpretation is a different point of view and each point of view enables us to see different and new aspects, or old aspects under a different and new light.

In what follows I propose an interpretation of this supposed old hat, hoping that it really might be different and thus highlight new aspects. In particular I will suggest considering the scientific theories as *hypothetical scientific representations*, i.e., conceptual tools made for cognitive aims, and the models as fictions, or, rather, as *fictive scientific representations*; that is, 'as if' or 'als ob' constructions, i.e., conceptual tools made for pragmatic aims. Starting from this point, that will be argued by resorting to some of the seminal works at the dawn of the contemporary philosophy of science, also a taxonomy of the models will be suggested.

2 Scientific theories as hypothetical scientific representations of the world

The interpretative proposal according to which scientific theories are hypothetical representations of the world finds its ancestor in the approach presented by two physicist-philosophers of the end of the XIX century, that is, H. Hertz and L. Boltzmann. For them, the scientific theories are to be thought of as *Bilder*.

In the classical English edition of Hertz's 1894 masterpiece – *Die Prinzipien der Mechanik* –, the term *'Bild'* was translated by 'image', but I believe that a more consistent and less ambiguous translation should be *'representation'*[1]. On the other hand, a *Bild*, for Hertz and Boltzmann, is something which allows us to intellectually orient ourselves in the world, or, rather, in a part of the world. Moreover, this 'something' has been objectively *built (gebildet)* by the knowing subject, according to certain rules. And the term *representation* catches exactly this bunch of senses, provided it is intended as a *construction, made in a certain language – in particular, in a mathematical language –, which presents objectively the hypothetical or fictive elements of the world and their hypothetical or fictive relations in order to know and live.*

2.1 Hertz's and Boltzmann's interpretations

In the great introduction to *Die Prinzipien der Mechanik*, Hertz, by taking up again Kant's 'Copernican revolution', develops an interpretation of the scientific theory according to which the latter is considered neither as the true description of the world, nor as the most truthlike one, but – as Boltzmann will claim later –

1 It should be noted that the term 'image' involves some semantical ambiguities since it has been used both in the sense of 'objective copy of what there is in the world' and in the sense of 'subjective copy of what there is world'. In both cases, the meaning is very far from what will be proposed here.

> as a mere picture [i.e. representation – *Bild* –][2] of nature [...] which at the present moment allows one to give the most uniform and comprehensive account of the totality of phenomena (Boltzmann 1899, p. 83)[3].

In primis, this means considering the theories not as true or truth-like formalisations, but as attempts made by the knowing subject to grasp the empirical data. It follows that the scientist is neither he who is looking for the intrinsic structure of the world, nor he who is coming closer and closer to such an intrinsic structure. Instead, he is the constructor of representations which allow him to grasp the empirical facts and so give them cognitive significance.

However, according to Hertz and Boltzmann, the role of representations is not limited to record in a coherent and economic way what is already empirically known. Actually, the representations also have to predict future events.

Therefore, it is not surprising that Hertz opens his introduction emphasising this point:

> To draw inferences as to the future from the past, we adopt the following process. We form for ourselves images [representations] or symbols of external objects; and the form which we give them is such that the necessary consequents of the images [representations] in thought are always the images [representations] of the necessary consequents in nature of the things pictured. In order that this requirement may be satisfied, there must be a certain conformity between nature and our thought (Hertz 1894, p. 1).

Thus, the representations do not tell us how things are in themselves, but, Kantianly, how 'our conceptions of things' are (Id.).

Someone might object that if the representations are our constructions it is not clear how there can be strong relations between them and the empirical world. To answer, it is necessary to bear in mind the three criteria that, according to Hertz, allow us to choose the best representation *(but not the true or the most truth-like one)* among those we can construct[4].

2 Note that here the term *'Bild'* has been translated by 'picture'.
3 It should be noted that J.C. Maxwell had already proposed a similar interpretation.
4 Of course we are tackling the question of the *underdetermination of theories by data*.

The first criteria, *the requirement of permissibility*, states that only the theories which do not 'contradict the laws of thought' have to be considered. That is, there has to be coherence between the representations and the laws of thought.

Then, among the permissible theories, those which are not contradicted by experience have to be chosen, that is, those which satisfy *the requirement of empirical correctness*. On the other hand, Hertz is perfectly aware that the scientific representations are hypothetical constructions made to grasp reality and, therefore, that reality plays the main role in valuing them. Only by appealing to experience can they be confirmed or refuted:

> [...the representation] has the character of hypothesis which is accepted tentatively and awaits sudden refutation by a single example or gradual confirmation by a large number of examples (Ivi p. 36)[5].

Finally, among the theories that are both permissible and empirically correct we have to direct our attention towards the only one which satisfies *the requirements of appropriateness*. That is, the most distinct one, in the sense that it includes the greatest number of essential relations, and the simplest one, in the sense that, besides the essential relations, it contains the smallest number of superfluous and empty relations, even if

> empty relations cannot be altogether avoided: they enter into the images [representations] because they are simply images [representations], – images [representations] produced by our mind and necessarily affected by the characteristics of its mode of portrayal (Ivi p. 2).

Surely, the third requirement appears rather ambiguous and vague, as Hertz himself writes. Before determining the most appropriate representation we should determine what is really essential (as to the distinctness) and what is really less divergent from such an essentiality (as to the simplicity). However, also the two other requirements, espe-

5 It is worth noticing that this passage was published in the 1894, that is, 40 years before the German edition of 1934 Popper's *Logik der Forschung*, where the thesis of falsifiability and the asymmetry between verification and falsification were presented with great emphasis.

cially the first one, are not trivially interpretable and therefore easily usable for settling the underdetermination.

Let us leave aside this interpretative questions and make content ourselves with the fact that we have found two heavy fathers, i.e., Hertz and Boltzmann, who support the idea that the theories are (neither true nor truth-like) representations constructed both to hypothetically represent the phenomena in a comprehensive and homogeneous way and to hypothetically predict new ones.

It should be remarked that this approach[6] provides us with a good *antirealism on the theories* (but *not an antirealism on the entities*), which teaches us that a scientific theory is neither a mirror copy of reality nor its approximation but a hypothetical representation constructed by the knowing subject to know and live.

3 Models as fictive scientific representations of the world

Now let us turn to the models, which I consider as *fictive scientific representations*. However, it should be noted immediately that, according to some authors, interpreting scientific theories as representations means interpreting them as models; as, for example, the Hertzian Wittgenstein of the *Tractatus* proposes: 'Das Bild ist ein Modell der Wirklicheit', i.e., 'The picture [representation] is a model of reality'[7] (Wittgenstein 1921, § 2.12).

Already this remark should make us aware that the term 'model' is not always used in the same way. First of all, a model may be something whose structural and/or functional relations are something to aspire to, that is, it may be thought of as an ideal-type *à la* M. Weber. Secondly, it may be something whose structural and/or functional rela-

6 Actually, the two physicist-philosophers differ as to the epistemological status of the principles which support the theories. In particular, according to Hertz they are *a priori* but distinct from the laws of thought. While according to Boltzmann, they are the laws of thought and even if they appear to be not empirically refutable, actually they are. They are empirically testable and not yet falsified propositions which have succeeded in the cultural evolution.

7 Notice that in C.K. Ogden's English translation of the *Tractatus*, the term 'Bild' is translated by 'picture'.

tions have to be reproduced somehow (for example, the model of a statue is what the statue reproduces). Finally, a model may be something whose structural and/or functional relations reproduce somehow the structural and/or functional relations of something else. *Mutatis mutandis*, it is this third sense that interests us.

However let us come back to the idea that the scientific theory itself may be interpreted as a model[8]. In this way misunderstandings may arise. Boltzmann, in the entry 'model' of the *Encyclopaedia Britannica*, explains that 'purely abstract conceptions should be helped by objective and comprehensive models in cases where the mass of matter cannot be adequately dealt with directly' (Boltzmann 1902, p. 215). This means that to acquire knowledge we need, beyond the theories – which are general and abstract representations –, particular representations which allow us to better grasp a given set of empirical facts. Allowing for this situation and for the sake of clarity, we should distinguish the level of the general abstract representations (theories) from the level of the particular representations (models, in Boltzmann's sense). It follows that if we wish to continue interpreting the theory-representation as a model, we should take care to stress that it is a model of models, that is, a large representation containing smaller and sometime independent representations.

Therefore, to avoid possible misunderstandings arising from these different uses, in the following I will not refer the term 'model' to the theory, also to emphasise that the representations offered by the theories are epistemologically different from the representations offered by the models. The former are *hypothetical representations*, the latter are *fictive representations*, as I am going to show.

8 Note that, according to Duhem (1914), the theory is a model in the sense that it is an axiomatised system connected to the empirical world by means of correspondence laws. As well-known, this is also the core of the interpretation proposed by R. Carnap, C.G. Hempel, E. Nagel, and R. Braithwaite.

3.1 *As if the world were*

Speaking about fictive representations means coming back to the approach proposed in 1911 by H. Vaihinger in *Die Philosophie des Als Ob*, which had (and has) the sad fate of being more quoted than read. Many authors affirm that the German philosopher claims that 'all is fiction'. Unfortunately this is not true. Actually, according to Vaihinger not all is fiction, and not all fictions are similar.

His theses are not so trivial and his fictionalism is a direct consequence of his historical studies. He was a leading character of the *Kantphilologie* and, between 1881 and 1892, author of a huge *Commentar zu Kants Kritik der reinen Vernunft*. And the source of his fictionalism springs exactly from the pages of the *Critique* in which Kant states that the ideas of pure reason – that is, the ideas of 'soul', 'world' and 'God' – are 'as if' having the pragmatic (and non theoretical) aim of leading to research[9].

3.1.1 *Remembering Vaihinger*

> *Fictio* means, in the first place, an *activity* of *fingere*, that is to say, of constructing, forming, giving shape, abstracting, presenting artistically fashioning: conceiving, thinking, imaging, assuming, planning, devising, inventing. Secondly, it refers to the *product* of these activities, the fictional assumption, fabrication, creation, the imaginal case (Vaihinger 1911, p. 81).

In this way Vaihinger explains the meaning of the term 'fiction', that is, the core of his theory of knowledge.

According to him, to gain knowledge, the human being constructs mental representations of the worlds which are totally different from the actual world. That is, they do not mirror the actual world at all. However there are three classes of mental representations: *dogmas*, *hypotheses*, and *fictions*; but only the third ones have a relevant role, satisfying in a better way the aim they have been constructed for, since they allow us to orient ourselves in the world and, thus, to predict fu-

9 So far as I can tell, among the contemporary philosophers of science only Fine (1993) seems to be aware of the non-triviality of Vaihinger's proposal.

ture events. Since fictions have only pragmatic aims, they should not be considered as theoretical concepts, but, rather, as pragmatic concepts.

Therefore, there is a theoretical and epistemological gap between the actual world in which we want to orient ourselves, also predictively, and the fictions we construct to realise it. However, such a gap vanishes whenever the actual world is pragmatically replaced with the imaginary world represented by the fictions. Even if the fictions represent an as-if-world, they are a (secondary) product of the actual world in which they allow us to live: 'The representations are not at all pictures[10] of the becoming; yet they themselves are the becoming; rather, a part of the cosmic becoming' (Ivi, ch. XIII). Thus, the human being and his thought belong to the actual world whose imaginary representation is given by the fictions. Unavoidably, it follows that the aim of thought is not thought itself but its fictive products that facilitate action and life.

Even if fictions are mere imaginary constructs different from the actual world, i.e., pure as-if-constructs, it should not be surprising, from an evolutionary point of view, that they satisfy in such an excellent way their pragmatic aims. In fact, during his long history, the human being has been constructing fictions which have increasingly satisfied his necessity to act and live. That is, if the core of fictionalism is given by Kantian representationalism, totally freed from apodicticy and transported to pragmatic shores by means of the emphasis on the status of transcendental ideas, Vaihinger, in order to justify the pragmatical success of the imaginary concepts, has recourse to a typical evolutionary approach, even if coloured in a teleological way.

It is necessary to underline that, according to Vaihinger, fictions, and therefore fictive judgements, are completely different from hypotheses, and therefore from hypothetical judgements. Hypotheses are mental representations which, *conjecturally,* tell us *how the world is.* Instead, fictions are mental imaginary representations by means of which *we postulate an as-if-world.* Therefore, hypotheses have an aim which is epistemologically different from the aim of the fictions. Being attempts to representatively grasp the actual *world without excep-*

10 Here 'picture' means 'true copy'.

tions, hypotheses have a cognitive and theoretical aim. Instead, as stated above, fictions have pragmatic aims, being instruments to allows us to act and live in the actual world, of which they are *an exceptional representation.*

This means that while the hypotheses may be refuted when contradicted by the world they try to grasp[11], the fictions cannot, being, in a certain sense, contradicted by the actual world from the very beginning.

Therefore, both of them are representations, but the hypotheses are intentionally not constructed against the facts, while the fictions are intentionally constructed against the facts. However, we can deduce consequences also from fictions. If the former do not match the world, the latter have to be rejected. Yet this rejection is not made for theoretical and cognitive reasons (as happens in the case of hypotheses) but for pragmatic reasons: *an instrument is simply shelved when it becomes useless for the aim for which it has been constructed* (Ivi, ch. XXI). Analogously, while we are theoretically and cognitively justified in maintaining a hypothesis confirmed by data, the only justification in maintaining a fiction is given by the pragmatic 'utility it gives to the empirical science' (Id.).

4 Theories and models

At this point, I may put together what has been said about hypothetical representations (theories) and fictive representations (models).

Prima facie, scientific theories are hypothetical representations which allow us

1. to organise in a general, abstract and formal way the empirical facts we already know;
2. to predict, at least statistically, new empirical facts;
3. to give cognitive significance to the old and new empirical facts.

Nevertheless, if we really want to arrive at the empirical facts we have to focus such general and abstract theories to the specific empirical

11 Note that around 20 years after, Popper will claim the same thing.

situation we want to analyse. This focusing is facilitated by a particular kind of models by means of which we consider only those aspects of the analysed situation that we conjecture to be necessary to successfully deal with it. That is, we construct a fictive representation of it by focusing the mother-theory, i.e., the hypothetical representation. And it is exactly by means of such a fictive representation of an as-if-world that we link the hypothetical representations with the empirical facts.

However, it may be the case that the problem to be solved is too complex if tackled starting from a hypothetical representation, or it seems impossible that the thought succeeds in disentangling the facts under analysis. In these situations, we replace such a hypothetical representation with a fictive representation that makes the problem tractable, i.e., in this case we use a kind of model different from the previous one.

A third type of models enters the scene when there are facts which cannot be arranged in any hypothetical representation we have and, at the same time, we are not able to propose a new one. Therefore, if we want to organise them somehow, we have to construct a fictive phenomenological representation.

In all these cases, fictions are not tentatively proposed as theoretical tools to grasp the actual world but as pragmatic tools to solve specific problems and/or to permit the link between the general and abstract level of theories and the level of the empirical facts.

All of this can be schematised in the following way:

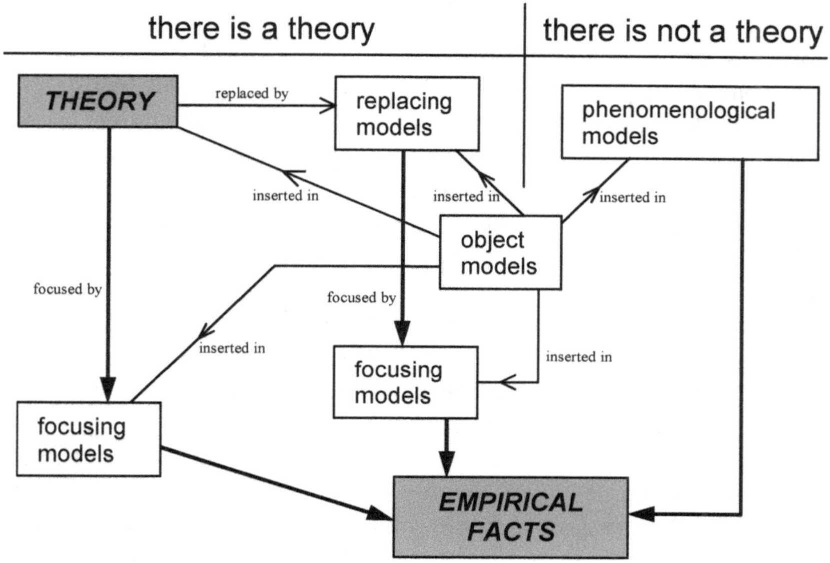

Before discussing and exemplifying these possibilities, two points are worth underlining.

1. Whenever I affirm that a theory represents hypothetically *how the world is*, while a model represents an *as-if-world*, implicitly I claim that there is an analogy between what is represented, respectively, by the theory and the model and the world. In both cases a comparison exists between the constructed structures and/or functions and the actual structures and/or functions. However, the comparison is epistemologically different in the two situations. As Vaihinger underlines (Ivi, ch. IV and XXII), in the first case we have a *hypothetical analogy* between the hypothetical representation and the actual world, in the second case we have a *fictive analogy* between the as-if-world and the actual world[12].

12 It is not my intention to analyse the general notion of scientific analogy. I shall limit myself to discuss the static/structural role of the fictive analogies inside the structure of the scientific knowledge. It follows that I will discuss neither the dynamic aspect of the models, nor their explanatory role. On the other hand, these are the main topics of the analogical interpretation of the models proposed by

2. Working with models may be surprising because from a fiction, which we know to be different from the actual world, it is possible to infer conclusions which are in good agreement with such a world.

An emblematic example is given by the model of the universe proposed, within General Relativity, by K. Schwarzchild in 1916. The fictive world represented by this model is very different from the usual conception of the world around us. It talks about a world composed of only a non-rotating spherical body and nothing else. Nevertheless, in such a model, which evidently represents a fiction, things work so well that the correct values for the three classical tests of General Relativity are obtained (the gravitational red shift, the deflection of the trajectory of a photon approaching the source of a gravitational field, the perihelion precession of Mercury).

That is, if the empirical tests were carried out (and they were actually carried out), one would find (and it was found) that *the empirical values are 'more or less' the ones found as consequences of Schwarzchild's fictive world*. How is it possible that predictions deduced from a model representing a completely fictive world are in close agreement with the empirical values found by measuring the actual world?

All of this might seem strange and mysterious. Yet, even if the model represents a fictive world, the fiction is not at all fanciful or arbitrary. It is constructed by taking into account only certain aspects of the actual world and neglecting others. But these neglected aspects are completely irrelevant as regards the predictions and their empirical fulfilment[13]. It is obvious that the actual universe is not made up of

---

Black (1962) and Hesse (1966). It should be noted that Boltzmann (1902) already talked about models as analogies.

13 There is a sort of necessary heuristic principle which allows us to construct theories and models: the principle of the fictive separability of the world, according to which one considers only and uniquely those aspects of the actual world which are necessary to solve that specific theoretical or pragmatical problem; all the rest is considered insignificant. It is necessary, because without it the scientific representations cannot be made. It is heuristic, because it leads towards the construction of the scientific representations. It is fictive, because one acts as if the world were in a certain way knowing that it is not so. This principle is different from the clause ceteris paribus. If the principle of the fictive separability of the world allows us to construct scientific representations by considering only what is important for given theoretical or pragmatical aims, the clause ceteris paribus allows us

only a non-rotating spherical Sun. Nevertheless, to add other elements would be completely irrelevant as to the problem. Certainly, there is a gap between the theoretical values derived from the model and the experimental values derived from the empirical measurements. They are only 'more or less' the same. *But such a gap is totally absorbed by the play of the experimental errors*. This is relevant: *the pragmatical validity of a model is discussed considering its empirical confidence level, that is, to what extent the numerical values it allows match the empirical values we actually find by means of the measurement apparatuses*. But the empirical values are obtained within an error interval which is not only due to the imperfection of the measurement apparatus but also to the perturbations caused by the rest of the actual world, that is, the part of the world not considered by the model. Yet these perturbations are irrelevant as to the very problem tackled by the model.

The initial surprise should have vanished now. The conclusions derived from a model, that is, from a fiction, are in close agreement with the empirical observations because, and as long as, the error interval of the measurements absorbs the effects due to aspects of the actual world neglected by the model.

4.1 A taxonomy of the models

Following what said above, we may divide the models in three main classes: 1) *principle models*, 2) *phenomenological models* and 3) *object models*.

---

a particular idea of the epistemological validity of the scientific representations. With reference to the fact that 'Le savants n'ont jamais méconnu cette vérite', that is, that the scientific representations are valid only ceteris paribus, cf. Poincaré (1902). As known, the principle of fictive separability has to do with what G. Galilei stated: *'quando il filosofo geometra vuol riconoscere in concreto gli effetti dimostrati in astratto, bisogna che difalchi gli impedimenti della materia; che se ciò saprà fare, io vi assicuro che le cose si riscontreranno non meno aggiustatamente che i computi aritmetici'* (Galilei 1632, p. 234).

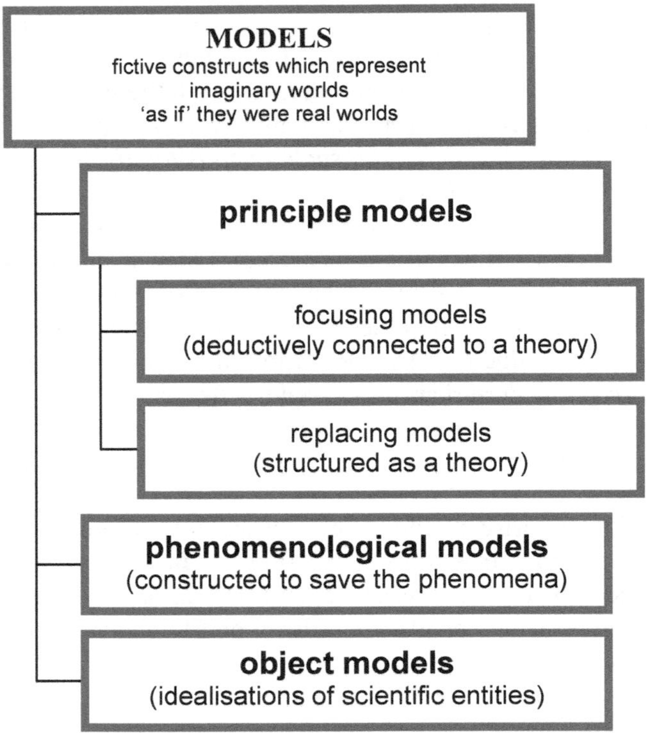

4.1.1 Principle models

By *principle models* I mean the models which, owing to the maintenance of the characteristic structure of theories (that is, principles and consequences), seem to offer a hypothetical representation equal to the one offered by the theories. Actually, *they offer only a simulacrum account of the hypothetical representation*: being models, the world they represent is not the real world, but an as-if-world which fictively, and therefore pragmatically, is considered instead of the actual one.

These models are the best known, but it is worth saying something about them. First of all, it seems that they belong in particular to physical sciences, where, differently from other scientific domains such as the bio-medical one or the chemical one, there are well-structured, from a logical point of view, theories. Secondly, they may be sub-divided in 1) models which focus a theory otherwise not able to grasp particular empirical situations *(focusing models)*; 2) models which replace an entire theory whenever it is impossible to utilise it owing to the complexity of calculus it involves *(replacing models)*.

4.1.2 Focusing models

In this case, on the one hand, we are dealing with models which allow us to grasp a particular empirical situation by focusing the general and abstract theory-representation they are deduced from. On the other hand, these models enable the link between such a theory and the empirical level. Of course they do not contradict the hypothetical representation they are deduced from: their premises are directly derived from the premises of the correlated hypothetical representation by imposing certain constraints which, on the one hand, focus them on the issue in question and, on the other hand, idealise the situation. Yet, although they are, in this sense, coherent with their mother-theory, they contradict our expectations of how the facts of the world are: it is the process of focusing and idealisation that leads to the construction of a fiction.

The focusing models may be exemplified in an emblematic way by means of General Relativity. The mathematical structure of General Relativity is a differential manifold with a Riemannian locally pseudo-Euclidean metric. Assigning a metric to the space-time structure means assigning peculiar relations among its points. Therefore, studying the explicit form of the metric tensor is equivalent to studying the gravitational field. Such an explicit form is obtained by solving the field equations, which are differential equations univocally solvable, by means of the Cauchy theorem, only by starting from well-determined initial conditions. It means that there are as many different solutions as different initial conditions and each particular solution gives a particular model of universe.

That is, on the one hand, there is the general and abstract representation of General Relativity which culminates in the field equations, but which is unable to represent specific universes. On the other hand, there are the focusing models, deduced from it, which are able to satisfy such a specific request. Therefore, each model derived from such a representation necessarily represents an *as-if-universe*, as the Schwarzchild model we have already seen.

4.1.3 Replacing models

The replacing models are constructed parallelly and analogically to the theories they replace. Therefore, they maintain unalterated the logical structure of the latter. However, representing a fictive world, they offer a simulacrum account of a representation of the actual world. That is, it seems they represent the real world, but actually they deal with an imaginary world which is fictively analogous to the world represented by the replaced theory. These models are used when it is too complex to handle the theory in a satisfactory way. This means that their premises not only go against the facts (because they are models), but also against what is claimed by the premises of the replaced theory. This is the main difference with the case of focusing models.

A classical example of the replacing models is given by the Kinetic Theory of Gas. We could have a physical representation describing a system of gas. Nevertheless, it is not easily tractable for predictive aims. Thus, we make up a first fiction considering the gas molecules as rigid spheres whose mean distance is greater than their diameters and whose elastic collisions are governed by the laws of Classical Mechanics. At this step, the fiction is not very good, because the situation is still non-tractable[14]. To overcome this obstacle we have to add, as J.C. Maxwell did, to the initial fiction three statistical assumptions, i.e., three new fictions: 1) the distribution of particles is uniform; 2) each component of the velocity is independent of the other ones;

14 It is sufficient to think that a mole of gas contains $6x10^{23}$ (Avogadro's number) particles. This means that in order to find the equation of the motion for such a system, $1.8x10^{24}$ equations should be solved by means of $3.6x10^{24}$ initial conditions. Even if we could solve this enormous system, we should sum the $3.6x10^{24}$ solutions to obtain the value of the macroscopic variables which we are interested in.

3) the velocity probability distribution is isotropic. By means of these three assumptions, it is possible to construct an easily tractable model that can predict physical situations.

At the beginning, we had a perfectly deterministic physical situation which could be represented in a perfectly deterministic way by a representation. But this representation would have been non-tractable owing to the enormous number of variables in play. Therefore, we were forced to construct a mathematised model which represented a fictive physical world.

Usually, a replacing model is also particularised by means of the focusing models, which enable the link between the level of the general and abstract assumptions of the replacing model and the level of the empirical facts[15].

4.2 Phenomenological models

While principle models maintain the typical structure of a theory, so that they may be thought of as a simulacrum account of a representation of the actual world, phenomenological models do not have such a structure at all. They might be considered as a sort of spider's web, constructed in order to capture flies, i.e., observable facts that, otherwise, would be without any correlation. In this case, the scientist has neither a hypothetical representation for grasping the facts, nor a principle model capable of organising them. Therefore, he is forced to cre-

15 It should be noted that constructing a mathematical model of a physical theory is completely different from constructing a replacing model of it. Under the latter process there is the notion of material analogy, as clarified by Hesse (1966); under the former there is the notion of formal analogy, as stressed by Redhead (1980). Here, we have an isomorphism between the physical theory and the mathematical theory by means of which the first is 'dressed again'. Redhead (1975) and (1980) points out that the new 'dress' can be larger than the older one. In this case, there is what he calls a '"surplus" structure in the mathematical formulation of a physical theory'. On the other hand, a mathematical 'model' of a physical theory cannot be considered a model in the sense discussed here: it is not a fiction, not an as if. We do not feign anything when we reformulate a theory by means of new mathematics.

ate, practically from nothing, a model. He borrows mathematical and non-mathematical pieces from here and there trying to put them together in order to construct something that allows him to correlate the data. Then, he constructs a fiction, rather a phenomenological model, that he hopes is capable of *saving the phenomena.*

In this way, I am suggesting that the only real case in which a scientist *saves the phenomena* happens when he is working with phenomenological models[16]. These have no purpose other than being a good patch which covers a theoretical incapacity to find deeper explanations by means of fundamental laws. They are fictions of the highest degree which have no significance besides being a good syntactical web to catch empirical phenomena.

Let us consider, for example, the model proposed by Lord Rayleigh and J.M. Jeans, between 1900 and 1905, to solve the problem of the black body. The two scientists considered the black body *as if* it were a cavity whose walls emitted electromagnetic radiation with frequency depending on their temperature. In order to construct the model, they borrowed results belonging to other theories. From the Classical Electromagnetic Theory, they took the fact that the radiation inside a cavity had to have the form of stationary waves with nodes at the walls. Geometrically, they derived the number of standing waves within a certain frequency range. Finally, they used the Kinetic Theory to calculate the average total energy of the waves of the system in thermal equilibrium. By means of these 'pieces' they arrived at an expression for the energy density. That is, they constructed a phenomenological model by means of which they hoped to have saved the phenomenon of black body radiation.

By the way, it is known that their conclusion did not match with the experimental values. Actually, there was an agreement only for low frequencies, but this was insufficient to consider the model positively. Thus Rayleigh and Jeans' model was abandoned: the pragmatic aims for which it was constructed were not satisfied. As said before, once its utility is exhausted, the model is simply put aside.

16 Of course, the matter changes when an entire theory is interpreted instrumentalistically.

In physical sciences there is a great use of phenomenological models, as is well exemplified by theoretical Nuclear Physics which may be thought of as composed exclusively of phenomenological models (cf. Boniolo, Petrovich, Pisent 2002). Moreover also all the mathematical models used in the bio-medical sciences, such as the so-called 'Hardy-Weinberg law' in population genetics or the Turing model for the morphogenesis, seem to be nothing but phenomenological models.

Let us consider, for example, the model describing the variation of a bacteria population. Let $N(t)$ be the number of bacteria at the time t and $N(t+\Delta t)$ the number of bacteria after a time interval Δt. That is, at the time $(t+\Delta t)$ the number is $N(t+\Delta t)=N(t)+\Delta N$. Of course the variation of the population is given by the bacteria which are born minus the bacteria which are dead. If B is the natality constant and D the mortality constant, after Δt we have a population equal to $N=[B\cdot N(t)\ \Delta t-D\cdot N(t)\ \Delta t]=(B-D)N(t)\ \Delta t$. It means that in the interval Δt there is a variation equal to $\Delta N/\Delta t=(B-D)N(t)$. This equation, for $\Delta t\rightarrow 0$, becomes $dN/dt=(B-D)N$. That is, we have $dN/N=(B-D)dt$. The solution of this differential equation is $N=e^C\ e^{(B-D)t}$, where C is the integration constant. If we call $N_0=e^C$ the number of bacteria at the initial time t_0, then the variation of bacteria population is described by $N=N_0\ e^{(B-D)t}$.

All the above mentioned quantities, i.e., N_0, B and D, have a strictly phenomenological significance, in the sense that they have been obtained by empirical and statistical analysis of the population. There is no theory from which the equation of the variation is obtained. It has been constructed, in a very simple way and by using trivial mathematics, only *to save the phenomenon* of the variation of the population; nothing more[17].

17 If a phenomenological model is constructed only to save the phenomena, it means that it correlates the facts but it cannot explain them nomologically. At most we have, as Cartwright (1983) correctly proposes, a 'simulacrum account of explanation'. That is, we have something which has the form of the explanation but which has not its substance. Actually, each class of models allows us a 'simulacrum account of explanation'. However, each class offers a different kind of 'simulacrum', and this fact should be taken into consideration.

4.3 Object models

In a certain sense, object models are atypical because, unlike models of the principles and phenomenological models, they do not have empirically testable consequences. Instead, they enable the theories, or the models, which they are inserted in, to have consequences empirically checkable. They are nothing but fictive constructions by means of which we idealise real objects. By means of them, *we make regular* (that is, tractable by *regulae*) *the non-regular objects of the actual world*. In other words, practising science also means, as Galilei taught us, *'difalcare gli impedimenti'* (getting rid of impediments), that is, replacing the actual 'dirty' objects with clean and more easily (sometimes more mathematically) handable fictions such as the rigid body, the point mass or the point charge, the spherical distribution, the ideal population, etc.

In this way, two relevant results are obtained. On the one hand, we simplify, sometimes drastically, the analysis we are performing inside a theory, or a model. On the other hand, we arrive at one of the main concerns of sciences: *the objects*.

It does not matter if these objects are idealised and fictive. At this point, what I have said about the match between the consequences of a model (i.e., the consequences of a fiction) and the empirical facts should be recalled. There is nothing strange in the fact that science, in spite of dealing with idealised objects, is able to successfully represent actual objects. Again, what enables such success is the play of the experimental errors, which are necessarily connected with each measurement and observation. We can use a *particular* object model (e.g., the fiction of the massive point, or of the spherical mass distribution) of that *particular* actual object (e.g., the Sun) *because* in that *particular* problem the divergence between fiction and reality is totally absorbed by the empirical errors. *The secret of the success of science lies exactly in the interval of the experimental and observational errors*.

I will briefly discuss a paradigmatic object fiction, that is, the continuous charge distribution, because it allows me to underline the limits inside which an object model can be applied.

Using the continuous charge distribution means introducing a scalar function: the charge density $\sigma(x,y,z)$. It is obvious that this is a fiction: the source of the electric field is not a sort of charged magma but a system of charges with finite, even if extremely small, dimensions. The convenience of using this fiction is evident when we have to tackle great systems, but it becomes clearer when we realise that, by means of this object model, one is able to solve the problems connected with the presence of certain mathematical singularities. In fact, near an actual charge the field $E = \sum_j (q_j r_j)/r^2 \to \infty$ as $(1/r^2)$. This obstacle can be overcome either by considering that it is physically non-significant to calculate the field inside the charge, or by adopting the fiction of the continuous distribution. In this case, the field $E = \int (\sigma(x', y', z')/r^2) dx'dy'dz'$ does not go any longer to infinity, not even inside the distribution where the volume element goes as r^2dr, for $r \to 0$.

However, this fiction, like the other object fictions, can be conveniently used only in certain situations. For example, when the system is thought of *as if* it were sufficiently great to consider the element of volume *dxdydz* sufficiently small, but containing a suitable number of charges such that the charge of the volume is $\sigma dxdydz$. Moreover, the fiction allows also for some intellectual adjustment. For example, the Poisson equation ($\nabla^2\Phi = -4\pi\sigma$) is a differential equation that should be valid only locally, being the relation between the charge density in a point and the potential in an infinitesimal interval around this point. But we know that the continuous charge distribution is an object fiction. This fact leads to interpreting only macroscopically the Poisson equation and considering the charge distribution as an average value in a small, but sufficiently large to contain many elementary charges, region.

In the case just seen, we have a fictive object that can be used in certain phenomenological or theoretical situations. But we have also the case in which a phenomenological model works only for certain fictive objects, as happens for the model of the so-called 'Hardy-Weinberg law'. In few words, the 'Hardy-Weinberg law' state that in a given population the frequency of two alleles is constant if the population in question is a fictive population 1) great enough to be treated in a statistical way and characterised by 2) absence of genic mutation, 3) absence of sexual preferences, 4) absence of genic flux, 5) absence of natural selection. Of course this population is a fictive object, that is, an object model, in this case a population model.

5 Models and mathematical schemes

To avoid misunderstandings, when I talked about the replacing models I emphasised in a footnote the differences between this class of models and the mathematical reformulations of a theory. Another source of possible misunderstandings is given by the differential and integral equations of physical and bio-medical sciences. It is true that these equations are used to represent many physical situations (from Fluidodynamics to Celestial Mechanics, from Electrodynamics to Quantum Mechanics, etc) and bio-medical situations (from Population Genetics to morphogenesis, etc.). But they, considered as such, are something different from both a physical fiction and a bio-medical fiction, that is, from what characterises a *physical model* or a *bio-medical model*. Rather, they are *mathematical schemes*, i.e., abstract and void mathematical tools which acquire physical significance or bio-medical significance only when they are applied to specific physical or bio-medical situations. Without this 'something more', they are nothing but 'pieces of mathematics'. Historically, a mathematical scheme may be constructed strictly correlated with a model, but its usefulness may have overflowed into other fields becoming a sort of routine, a sort of *prêt-à-formaliser*, or *prêt-à-résourdre* analogous problems[18].

18 Also the Lagrangian and the Hamiltonian formalisms should be interpreted in the same way, i.e., as mathematical schemes. Thus, rigorously speaking, it is not correct to talk about the Hamiltonian model of the harmonic oscillator, or the Hamiltonian model of the hydrogen atom, and so on. This way of speaking conceals a misunderstanding because, actually, we are facing with the application of a mathematical scheme to formalise idealised physical situations. It follows that one should talk about the Hamiltonian of the model of the harmonic oscillator and the Hamiltonian of the model of the hydrogen atom, etc. It is the harmonic oscillator which is a model, not its Hamiltonian which is the particularisation of the general scheme of the Hamiltonian $(H(p,q,t)=\sum_i p_i q_i - L)$.

6 Models and mathematised theories

Another remark concerns the differences between mathematised representations and mathematised models. This is important since it allows us to put into a correct philosophical frame the mathematical formulations dealing with biology, economics, meteorology, sociology, the flux of traffic in a great city, the project of a car or a ship, etc.

Let us consider the case of the Theory of Deterministic Chaos (TDC). In this situation at the beginning we have a completely deterministic problem. However, now the initial conditions have a crucial role to play. It is sufficient that we know these conditions only with the slightest lack of precision and it becomes impossible to determine future states. From a well determined starting point, the system becomes soon chaotic and unpredictable. Nevertheless, even if the system is chaotic in ordinary Euclidean space, matters are different in the phase space. Here 'strange attractors' appear and allow us to find again, even if qualitatively, the lost predictiveness and the lost order.

As shown, in the case of the Kinetic Theory we have a physical system fictively represented by a replacing model which allows us to obtain predictions, although statistically. However, in the case of TDC, we are facing physical systems which, before the works by H.J. Poincaré and others, were not described by any hypothetical representation. Thus, the TDC is not a mathematised model of physical situations such as turbulence, crazed compass needle, etc., but it is a hypothetical representation from which it is possible to deduce idealised and fictive representations of aspects of the world, such as the smoke of a cigarette, the flux of water from the tap, the vortexes in the river, meteorological states, etc. Summing up, the representations of the latter phenomena are mathematised focusing models[19], but the TDC as a whole is a hypothetical representation, a conjecture, written in a mathematical way, on a part of the world.

At this point, we possess the philosophical tools to understand what it means to construct the mathematical model of a specific situa-

19 Morton (1993) calls these models 'mediating models', i.e., models which mediate between a 'governing theory' and the phenomena which the theory cannot be applied to, due to the complexity of the matter.

tion, e.g., a meteorological situation or a biological situation, in order to organise the data and possibly to achieve the prediction of future states. We have two cases:

1 we may deduce the model from a theory (of course if we have it, as happens in the case of the TDC);
2 we may construct the model entirely independently of a specific representation, but borrowing the necessary mathematical or non-mathematical pieces from here and there.

In both cases, we arrive at a mathematised model of a fictive situation, but in the first case we have a focusing model, whereas in the second case we have a phenomenological model.

Now we should grasp the difference, stressed also by Black (1962), between mathematised representations and mathematised models: the TDC is a mathematised (hypothetical) representation, i.e., a theory. But the formal representations of the economic cycles, of the traffic at an important crossroads, of the tension borne by a bridge, etc., are either mathematised focusing models (if there is the representation which they are deduced from), or mathematised phenomenological models (if there is not such a representation)[20].

20 At this point, the case of the Thermodynamics of Irreversible Processes (TIP) and the Theory of Catastrophes (TC) is worth considering. Already in the forties, I. Prigogine began to study what happens in open, non-linear and far from equilibrium chemical-physical systems. He reached a complete representation of these systems such that, even if their internal entropy increases according to the second principle of Thermodynamics, the variation of the total entropy may vanish, provided that there is a flux of negative entropy from the outside. This fact permits the creation of ordered and self-organised situations in cases not considered by Classical Thermodynamics. Of course, TIP has to be thought of as a theory as much as Classical Mechanics and Special Relativity, and not as a model. At a certain point, Prigogine became aware that a huge quantity of systems might be thought of as open, non-linear and far from equilibrium systems. Therefore, he expanded what had been found for the chemical-physical systems to these new situations. In such a way he thought he had found *a new alliance* (cf. Prigogine-Stengers (1979) between the world of nature and the world of life. By means of this approach, he was able to represent almost everything: from Bénard vortexes to Zhabotinski reaction, from the chemical reactions among molecules to the organisation of a colony of termites, from the growth and falling of an ecosystem to pre-biological evolution, to social-cultural development, and so on. Wherever

Bibliography

M. Black, 'Models and archetypes' in M. Black, *Model and metaphors. Studies in language and philosophy*, Cornell University Press, Ithaca (NY) 1962, pp. 219-243.

L. Boltzmann, 'Über die Entwicklung der Methoden der theoretischen Physik in neueren Zeit', 1899, now in L. Boltzmann, *Populäre Schriften*, 1905, Engl. transl., in L. Boltzmann, *Theoretical physics and philosphical problems*, Reidel, Dordrecht 1974, pp. 1-198.

L. Boltzmann, 'Models', Encyclopaedia Britannica, 1902, now in L. Boltzmann, *Theoretical physics and philosphical problems*, Reidel, Dordrecht 1974, pp. 211-220.

G. Boniolo, C. Petrovich, G. Pisent, 'Notes On The Philosophical Status Of Nuclear Physics' in *Foundations of Science*, 7, 2002, pp. 425-452.

N. Cartwright, *How the laws of physics lie*, Oxford University Press, Oxford 1983.

P. Duhem, *La théorie physique: son objet et sa structure*, 1914, Engl. transl., *The Aim and Structure of Physical Theory*, Princeton University Press, Princeton 1954.

A. Fine, 'Fictionalism' in *Midwest Studies in Philosophy*, 18, 1993, pp. 1-18.

---

there was the possibility of representing something as an open system, there Prigogine's approach struck. Omitting the *hybris* of such a holism, attention should be paid to the fact that the models for the cases above listed are focusing models because they are derived from a theory, the TIP. (An almost analogous situation is given by Haken (1981)'s 'synergetics'). Differently from TIP, TC, proposed by R. Thom, arose as a mathematical theory describing the drastic passage of a system from an evolution governed by a certain set of differential equations to an evolution governed by another set of differential equations. TC was also the object of many discussions, and many people have considered it as a sort of formal panacea able to describe an incredible quantity of physical, biological, social, historical, etc., phenomena. Soon TC has become a representation for every kind of transition of the world. In this case there is a mathematised abstract and powerful scheme which permits us the construction of mathematised phenomenological models able to fictively represent as-if-worlds, where there are physical bifurcations, revolts in the jails, biological mutations, historical or cultural revolutions, and so on.

G. Galilei, *Dialogo sopra i due massimi sistemi del mondo*, 1632 in G. Galilei, *Le opere*, Giunti Barbera, Firenze 1968, vol. VII.

H. Haken, *Erfolgsgeheimnisse der Natur*, Deutsche Verlags-Anstalt, Stuttgart 1981.

H. Hertz, *Die Prinzipien der Mechanik*, 1984, Engl. transl., *The principles of mechanics*, Dover, New York 1956.

M. Hesse, *Models and analogies in science*, University of Notre Dame Press, Notre Dame (Ind) 1966.

A. Morton, 'Mathematical models: questions of trustworthiness', *Brit. J. Phil. Sci.*, 44, 1993, pp. 659-674.

H.J. Poincaré, *La science et l'hypothèse*, 1902-1907[2], Engl. transl., *Science and Hypothesis*, Dover, New York 1952.

I. Prigogine, I. Stengers, *La nouvelle alliance*, 1979, Engl. transl., *Order out of Chaos*, Bantan Books, New York 1984.

M. Redhead, 'Symmetry in intertheory relations', *Synthese*, 32, 1975, pp. 77-112.

M. Redhead, 'Models in physics', *Brit. J. Phil. Sci.*, 31, 1980, pp. 145-163.

H. Vaihinger, *Die Philosophie des 'als ob'*, 1911, Engl. transl., *The Philosophy of 'As If'*, Barnes and Noble, New York 1968.

L. Wittgenstein, *Tractatus Logico-philosophicus*, 1921, Routledge, London 1996.

Giovanni Boniolo, Dipartimento di Filosofia, Università di Padova

Mario Borillo

Technological risks or risks of reason? A reflection on cognitive technologies

Abstract:

The critique of the societal implications of technoscience cannot be conducted solely from the exterior, as it were, through the analysis of its manifest impact on the world. An in-depth understanding of its nature, and in particular of its underlying rationality, is also indispensable for shedding light on the radicality of the upheavals it provokes. Because it is in the intimacy of the meeting between computation and cognition that the operativity of reason is today concentrated, it is there, perhaps, that we might find one of the Gordian knots of the age. The discussion of this hypothesis will be articulated according to two sections whose complementarity could itself be revealing.

In the first section, we recall how the 'nervous system' of all the technologies – whose multiform impact affects work, the economy, forms of social life, culture, war, and so on – consists of a set of hardware devices and symbolic systems that are nothing other than the operational implementation of fundamental knowledge on computation and cognition. *However, the same principle of rationality that animates data-processing systems is used today as a conceptual framework for a theory of human cognition. This essential scientific encounter, because it relates to the nature and the exploitation of reason, explains the universality of its technical derivations.*

In the second section, an attempt is made to put into perspective the consequences of a meeting of this importance, when the extension of the knowledge thus associated is implemented in an instrumentality whose nature enables it to be introduced into practically every human activity. Indeed, because technology incorporates, by amalgamating them into data-processing devices (computers), increasingly broader segments of forms and functions of cognition, and because this operational assimilation is probably destined to widen considerably, we cannot but wonder about the repercussions of this emergence, which could, in the long term, attain an anthropological dimension.

At the same time as the knowledge of our mental nature begins to receive a scientific statute, which displaces the old metaphysical problems on 'being' and on the nature of the mind, the impact of cognitive technologies, in their multiple dimensions, born of the intimate fusion between data processing and cognition, promises to be too serious to be anything other than the subject of rigorous attention.

Keywords:
computation and cognition, computational reason, cognitive technologies, new frontiers to the formal representation of man and society, art, culture, cognition and computation, socialisation of computational reason, societal impact

1 Strategies of computational reason

1.1 New functions for 'machines' engaged on new technological frontiers

Perhaps, when developing a reflection of this nature, it would be useful to anchor it in the examination of a class of exemplary systems of technological changes currently in progress. By the number and the extreme diversity of the physical, biological, and even psychological constitutive elements of space exploration systems and their interdependent environments (when one thinks of the crews of manned space-shuttle flights or of the controllers on the ground of these flights), and by the corollary complexity of the evaluation procedures and the decision-making processes that are involved in all the phases of the achievement of a mission, and also in the exploitation of its results, man's activity in extraterrestrial space constitutes one of the fields in which is revealed, in the clearest and most anticipatory manner, for the great post-industrial technological systems, the omnipresence and central *rôle* of three closely dependent cognitive functions. These are: (a) *interpretation*, giving a meaning to a state of the world in question, in a certain context and for certain ends, beginning with information delivered by sensors; (b) *reasoning*, that is to say, carrying out, on symbolic representations of the world in question, the logical operations most capable of indicating (c) some optimal *decision-making processes* that are likely to lead to the desired results.

However, the novelty is not in the importance of these functions – largely recognised since the beginnings of automation – but in the progressive transfer, which appears inescapable, of a subset of these essential functions: namely, those for which neither mathematical nor more widely formal models yet exist, and which, for this reason, have traditionally been dealt with by a human operator (by a driver or pilot or controller) – in short, by a human mental and sensorimotor device, which, through the facility of language, we may still refer to as an data-processing system. Alas, too many unhappy experiences have shown that, in making this transition – progressive in its operational methods, but radical in its major implications – we have often not drawn all of the appropriate conclusions, and failure – the actualisation of risk – has struck in the field of space exploration as elsewhere (e.g. accidents resulting from failures of automatic piloting systems used in large aeroplanes, and certain failures of the control of nuclear reactors, and so on).

It is not necessary to enter into an instrumental description of the evolution of space exploration in order to reveal that the problems attached to the above-mentioned functions have become increasingly crucial with regard to their implications, and also more difficult to solve conceptually and technically. As one might expect from the combined effect of the appearance of a need – born of the increasing awareness of the limits of human aptitudes in the face of the complexity of the processes being controlled[1] – and of the availability of a new means, rich in potential, with which to face that need – namely the extraordinary developments, both material and logical, in the possibilities of computation – the *rôle* reserved for data-processing systems in the design and control of missions has become increasingly important. The remarkable point is the diversification of this *rôle*, which comprises more and more new functions that we could qualify as 'cognitive', in addition to the classical computational (i.e. numerical and mathematical) functions to which the founding physicist tradition of space exploration tends to confine it. Nevertheless, the importance of calculation in the modelling of physical components has not dimin-

1 Constraints related to the effective presence of man in extra-planetary space, and the limits, or even the impossibility, of this presence when certain physical thresholds are crossed, raise, in comparable terms, the question of the transfer to machines of certain human sensorimotor and cognitive aptitudes.

ished in the least – indeed, quite the contrary. Simply, other functions appear, of a completely different nature, which also make use of considerable data-processing resources.

Whatever its limitations, the *Mars Pathfinder* mission, launched on July 4, 1997, is in this respect strongly eloquent, not only for the part devolved to *Sojourner*, the semi-autonomous robot deployed on the surface of Mars (whose behaviour was partially controlled by an on-board computer), but also, and in an exemplary way, for that which it inaugurates in the way of co-operation between man and computer in the adventurous pursuit of the experiment.

1.2 Computation and cognition: anatomy of a meeting

What, then, reveals this 'co-operation', beyond the rudimentary extent to which it was restricted in the *Mars Pathfinder* mission, if it is not that which is of most significance in the architecture and the functional organisation of any system conceived as a 'partner' for a human operator? The architecture of a distributed computer system (in the case of *Pathfinder*, a part of which was is in the robot, and a part on the ground) can be made to include, on the one hand, sense organs (sensors) and multimodal organs of communication (oral, written, aural, visual, kinaesthetic), which are all providers of information for the digital 'brain', and on the other hand, extremely diverse types of effector, charged with achieving the desired interventions in the physical environment in which the robot finds itself. For our purposes, it is the nature of that which occurs in the 'brain' – that is to say, in the on-board data-processing system, properly called, during the temporal evolution of its exchanges with the external world and with the operator – which is most deserving of attention.

We shall raise thus, as constitutive functions of what one could metaphorically call the total 'psychophysiology' of the system: (1) interpretation of the state of the world perceived, and of the messages exchanged with the operator, suitably matched to the objectives pursued; (2) representation of the result of this interpretation, in a formal language fitted to the various associated computers; (3) 'reasoning', through the execution of calculations of every nature, numerical and logical, to evaluate a situation or the result of an action, to manage the dialogue with the operator, and so on; and (4) the making of decisions, in co-operation with the operator, regarding the relevant actions, always in the context of the objectives assigned to the mission. This sequential subdivision is obviously a convenience for the sake of explanation. In the material and formal reality of the system, all these processes are linked at all levels, from the perception to the decision, and vice versa, by dynamic loops of interaction, a simplified explanation of which would greatly exceed the modesty of this paper. The study of such functions and their integrative architectures is by now the object of intense research in all the developed countries, and their (partial) applications are multiplying.

'To perceive', 'to interpret', 'to represent', 'to reason', 'to communicate', 'to co-operate', to which we must add 'to memorise' and 'to learn': that these words must be used today (because as yet there are no others), to designate functions entrusted to a hardware device and symbolic system conceived and realised by man but external to his corporality, should suffice to alert us to the importance of the upheavals of every nature – and consequently to the gravity of the risks – brought about by this *machina* (Latin: invention, machine; derived from Greek *mêkhanê*: art, trick), in whatever field of human activity it is, or will sooner or later be, inserted.

Certainly, our evocation will be deprived of the restrictions and uncertainties that scientific prudence would impose, but it would be erroneous, for the sake of reassurance or laziness, to see a pure scientific and technical Utopia there. It need hardly be mentioned that the meanings of the terms quoted above are not the same when applied to human beings as when applied to machines, if only for the reason, too often forgotten, that we still know very little about what such terms cover when applied to man. If each one of us possesses a distinctive personal usage of such terms, their *scientific* description is just beginning, and has become the subject of systematic research within a coherent theoretical framework only since the advent of the science of cognition, within about the last half-century (beginning with the foundational work of Turing, Von Neumann, and Minsky, for example). When these terms refer to machines in which the corresponding functions are implemented, it is established that they have a less equivocal significance, more specifically linked and reduced to a context, to a particular end, and also that these functions will never be extremely interlaced and interdependent as they are in man, the product of a long process of evolution.

It is also with this limitation in mind that the development of research continues, even in computer science. Whereas once there was talk of designing 'intelligent' machines[2] that would be completely and universally autonomous in their behaviour (a naïvely anthropomorphic belief that the majority of researchers thought clever to let the popularisers attribute to them), there is today more talk of 'co-operative' machines: that is to say, machines possessing specific and separate aptitudes, limited to certain functions, but functions that are complementary to those of human beings, in the pursuit of some objective or other, and therefore presenting a certain degree of compatibility with human beings in the sphere of representations and symbolic functions.

It is from the point of view of this complementarity that we might best shed light on the strategy, and perhaps the challenges, of the great scientific (and, inseparably, technical) revolution in which we are now engaged. In its most basic characteristics, the work in progress seems to us to be circumscribed in a space delimited by three principal axes of research, whose strong articulation and coherence may upset not only our knowledge of ourselves, but also, through their possible consequences, of the whole anthropic sphere. Let us take the risk of giving an evocation of it, necessary for our purposes, but too deprived of

2 The controversy on the chess-playing vexations of the unhappy Kasparov illustrate the inanity of a discussion obsessed with 'the intelligence of the machine'. The true question is to know how the co-operation within the IBM team was constituted: how the strategies defined by the grandmasters led to programs conceived by algorithm specialists, as well as possible exploitation of theoretical and architectural (material) progress of data processing. Without the system, it is probable that the associated grandmasters would not have beaten Kasparov; and without them, the best machine would probably not have succeeded either. The interesting question is to know what cognitive and formal problems were faced in the effective transfer of high-level chess-playing experience to the data-processing system, and what will arise from this new knowledge in the maturation of computer science.

nuances and restrictions to accurately reflect the lively scientific debate surrounding it.

1.3 A. Computational reason constructs a representation of man ...

A certain number of disciplines – particularly neuroscience, psychology, ergonomics, linguistics, semantics, the philosophy of language[3], and the philosophy mind[4] – are by now associated in a project whose purpose is to define on a conceptual and empirical level – and, for some, to describe formally – the way in which the above-mentioned functions are carried out in man. The project has an extraordinary integrative aim that endeavours to span an arc from the biological substrate to the symbolic structures of language and thought. It need hardly be said that the disproportion of such an ambition is tempered by the fact that research proceeds with severe restrictions on the delimitation of the issues actually dealt with. However, as we said, this ambition is extended – and this is crucial – to a second stage, which tries to provide a formal representation, a symbolic model of the observed phenomena. It should be stressed that this formalisation is thought to be within the general theoretical framework of cognition – which is itself inspired very precisely, as we said of the framework of computation, (i.e. the theory of automata, their architecture and their programming) – and that it is expressed using the languages of logic and theoretical data processing.

The clearest example is given by the theories of syntax and semantics of natural languages, from the pioneering work of Chomsky, Schützenberger[5] and Montague[6] (and that of the large family of their

3 For a presentation of the principal current developments, see *Introduction à la philosophie du langage*, Daniel Laurier, Mardaga, 1993.

4 A very clear discussion of the most significant positions of analytical insight may be found in *Introduction à la philosophie de l'esprit*, Pascal Engel, La Découverte, 1994.

5 The linguistico-cognitive philosophy of Noam Chomsky was revealed, in its various forms, in many works. We shall hold as essential *Le langage et la pensée*, Payot, 1972. Marco Schützenberger, who died in 1996, was a French mathematician and philosopher of science, and one of the founders of theoretical computer science. His collaboration with Chomsky, in the mid-nineteen-fifties, made it pos-

explosion of successors)[7]; but it is also true for a number of theories of learning, and for models of reasoning, among others[8]. The fertilisation is reciprocal. Thus, on this last point, it is important to recall that psychological, linguistic, philosophical and computer research has led more recently to the development and the implementation of new logics (modal and non-monotonic, for example)[9], for reasoning in new and relevant ways on facts, knowledge and beliefs, but also on norms, temporal evolution, exceptions, and so on. These new languages of logic are in marked contrast to the rigid non-expressivity too often given to classical logic, but they preserve its properties of rigour (descriptive adequacy, logical consistency, etc.), and of *calculability* (the 'dynamicity' of the models expressed by recursive procedures), which make them of both theoretical and operational interest. The consequences of this meeting are no less important at the philosophical level, but they are not clearly intelligible, and their range is not seriously appraisable other than within the framework of the analytical philosophy of mind.

There is an important restriction. It is necessary to underline that we sketch here the symbolic approach in the cognitive sciences, whose exemplary paradigm would be the description of the functioning of language and its relation to meaning. A very important complementary approach involves the study of the biological roots of cognitive functions. Centred on the brain, it attempts to describe at the neuronal level the biological structures and functions that are brought

---

sible to give a rigorous formal base to the syntactic theories of natural languages and to those of programming languages.

6 Richard Montague, philosopher and American logician, deceased 1970, proposed a truth-conditional and recursive approach (that is to say, theoretically suited to calculation) of the semantics of natural languages. An exposé is given in *Logic and language*, Fréderic Nef, Hermès Editions, 1988.

7 For example, B.H. Partee, A. ter Meulen, R.E. Wall, *Mathematical Methods in Linguistics*, Kluwer Academic Pub., 1990

8 For an illustration of this methodology in the humanities, see M. Borillo, *Informatique pour les Sciences de l'Homme; Limites de la formalisation du Raisonnement*, Mardaga, Brussels, 1986.

9 An excellent work for the presentation of new developments in logic is *Méthodes logiques pour les sciences cognitives*, edited by Jacques Dubucs and François Lepage, Editions Hermès, 1995.

into play in the exercise of cognition. This approach is, with good reason, called *sub-symbolic*. This research, like that carried out within the framework of the symbolic paradigm, rests on the need for a strong connection between, on the one hand, experimental protocols and results, and on the other, the formal framework in which these observations could be expressed and possibly simulated on machines. In fact, the requirement of expressive relevance leads to different types of modelling, because they are now to be expressed in the mathematics of continuity and of parallel-processing structures[10], such as, for example, the so-called 'formal neuron networks'. These models, implanted on digital machines, will enable those machines to recognise images and forms, to identify speakers, and so on. The same machines, equipped with these new programs, will therefore be able to simulate cognitive functions designated 'of low level', complementary to those of 'high level' mentioned above.

1.4 B. ... and also a representation of the world as this man perceives it

If the aptitudes of the cognitive subject can thus, each in turn, be described at various levels, and sometimes simulated by a machine, a discussion in some ways complementary to the preceding one, to which it is closely bound, naturally follows. It relates to the 'nature' of the world as it is 'constructed' by the cognitive subject, and which manifests itself through the various relations that it maintains with him – for example, the world with which he is put into interaction by his sensorimotor organs, and which is at the centre of his mental life even in the absence of a real sensation (one can 'see red' with one's eyes closed). The problem becomes, then, one of describing worlds as they are structured by the perception, action, language, and reasoning of common experience. In these worlds, in contrast to the world of the natural sciences as we understand them, the meaning of *colours*, for

10 One could consider foundational the work of McClelland J. & Rumelhart D., *Parallel Distributed processing: Explorations in the Microstructure of Cognition*, MIT Press, 1986

example, is no more than partially expressed in terms of wave frequencies and energy spectra; and the meanings of linguistic expressions of space and movement – think of 'Paul runs towards Alice' – are not reduced to the measurement of geometrical structures or cinematics. Intervening also, in this last case, for example, are the communication objectives of the speaker, and some information on the identity and the intentions of the spatial entities evoked in the sentence.

Curiously, if the natural sciences were constructed from the outset with a view to eliminating the subject, in attempting to provide descriptions independent of his subjectivity, the cognitive work in progress aims at reintroducing that same subject, by linking him to a description of the world as it would have to be represented, hypothetically, at the mental level, if it is to allow control of our current activities. This is demonstrated by the research in semantics already cited, for example (as we exchange, through language, information that is precise and consistent enough, in general, to be used successfully in the objective world), or by the use of a 'naïve physics' (e.g., our knowledge on the behaviour of liquids is generally sufficient to prevent us from overfilling bath-tubs), or by the spatial and spatio-temporal reasoning of common sense (e.g., we are in general able to use knowledge associated with topography, and 'stored' in our memory, to determine an acceptable path to follow), and so on.

This brief evocation suffices to indicate the diversity, novelty and difficulty of projects that require widely interdisciplinary teams. These projects open some foundational and methodological problems, as do the majority of issues relating to cognition. Their theoretical outcome, however, consists in the definition of formal models of these knowledge and common aptitudes, generally expressed in the language of non-classical logics and of computational procedures. It should be noted that we require, of course, formal models that not only satisfies the usual conditions of descriptive adequacy, but ones that also have satisfactory computational properties. These questions depend initially on fundamental research, but they also open up the prospect, in the long term, of mastering certain modes of interaction of an artificial system: for example, a robot (the integral organisation of 'brain' + sensors + effectors), interacting with its material environment by using representations that would be 'compatible' with those of a human operator, in such a way that co-operation would be possible between man and machine in the pursuit of certain objectives, some of which would be beyond the reach of man alone.

For example, the importance of these research projects, if realised, will become clear in the exploration of outer space, when the long duration of data-transmission would mean that the robots sent to other celestial bodies would need to be equipped with forms of autonomous behaviour, even if only partial, but based on models who are elements of the terrestrial operator. In fact, rare would be the human activities unaffected by such implementation. One can thus imagine how such models, when made operational, would contribute to the co-operation between human operators themselves, when unfavourable circumstances – be they material (e.g. inscription in different sensory universes) or intellectual (e.g. gaps in the information available) – could be partially compensated for by each participant, through intelligible representations of the partner's world. This 'partner world', as it were, may comprise and exploit concepts as subtle as the epistemic difference between facts, knowledge and beliefs (cf. our remarks on non-classical logics).

1.5 Provisory Epilogue

It remains to suggest how the 'hard core' of computational reason – i.e. computer science itself, in its multiple metamorphoses – enters into a singular process of transformation that leads *necessarily* to an extending and deepening of its societal impact; how its intrinsic progress, at the level of the material architecture of machines, and at that of the theory of computation and the software which implements it, is amplified by the progressive assimilation of the empirical and formal results concerning the cognitive phenomena mentioned above; and how, by incorporating them into new technical devices, this 'soft' metamorphosis of computation might potentially install the operativity of computational reason at the heart of almost all human activities. It is in these radical terms, we think, that the coming partnership which we are already sharing with these 'machines' is posed. The question of risk is enrooted the new and most intimate *usage* of reason.

2 To capture cognitive technologies in action

To try to understand the extraordinary impact of data processing on all forms of human activity, perhaps there is no better solution than to think of man himself, and to consider that with each stage of his anthropographical course – for each of his activities – it is no longer man alone who is in the position of an actor, and that it is necessary to attempt, albeit through a game of the imagination, to substitute for him a 'couple' of a radically new type, to form what by a misleading material analogy we continue to call a machine[11]. The first symptom of the new, the configuration of this assemblage, takes forms so various that they reveal the troubling universality of a concept. Under the most tangible of materialities – that of the physical effectors on a robot – the data-processing system can operate alone for tasks traditionally reserved for a man (e.g. assembling the parts of a machine). It can 'co-operate' with man in an existence that is at once material (revealed by sensory interfaces) and abstract (e.g. the result of 'calculations' of the machine), when its decision-making propositions are compared with those of the brain. In so doing, it can form with him a totally hybrid functional couple (as realised, for example, in the cockpits of aeroplanes); or perhaps – in a situation more revealing at the level of thought – the couple may even be virtual, in that the machine is not materially used in the committed action, yet everything happens as if it were. Even if a man operates alone, the analysis of the problem, the conception of the world that it raises, and the decisions that result from this, are largely determined by the principles already interiorised

11 The metaphor of an attachment would be more suitable, as it is true that what joins together both partners depends more on research into efficacy than on the satisfaction of a passion. However, there are some exceptions... on the part of man.

by this man, of a triumphant 'computational reason'[12]. In general, that which tends to be true of every man at work tends to be true of everybody, as has be seen from the 1950's, in an increasingly larger and more varied number of spheres of activity. If, from now on, it must be necessary (in situations of action, communication, reflection, and so on) to substitute man with the couple he forms with the machine, the consequence is that it is necessary also to think of the society of men as the society of these couples governed by computational reason. The accelerated construction of the planetary network, by potentially opening for each couple the possibility of interacting directly with all others, seals the concrete existence of this society.

Grammatical and semantic restriction: such an assertion is certainly not yet true today, in 2003. The future conditional would be a better convention. The objective is to attempt to draw possible conclusions from the current state of affairs that these consequences raise, as it proven, of the particular logic of the objects that occupy us; or, in an infinitely more risky way, of 'logics' that one can lend to society and history, or more probably still, to the unprecedented association of these fundamental modes of thought and action.

2.1 The question of the limits: In what ways would man be irreplaceable?

The pedestal on which the praxeological universality of the man-computer couple stands – the concept that, within the limits of our knowledge, defines it today, in effect – is that of *computation*. It is on the theory of computation that all data-processing systems rest. It is remarkable, however, that it also constitutes the formal framework that has made it possible to found a theory of human cognition, and to make a start on its description. From its earliest days, the computer seemed a powerful amplifier of our capacities of calculation, but it is now becoming more and more evident that the revolution is not only one of scale, but also one of nature. Only the deep similarities that

12 But is there still a decision-making process in the world of production, of management, or of war, for which the computer would be materially absent?

seem to exist, between the cognitive structures of man and the principles that control the operation of this universal instrument, make it possible to understand the fecundity of current research on language, reasoning, perception, memorising, learning, and so on. This new knowledge indicates how the majority of our intellectual faculties could interact with computer systems, and find themselves potentially modified by this interaction.

One can think that whatever the nature and objectives of our actions, whatever the situations in which they are circumscribed, their operational rationality should, if the cognitivist hypotheses is verified, show more and more of the characteristic attributes of the computational world. The recourse to symbolic notation and to effective computation may proliferate in virtue of two fundamental assumptions of cognitivism: (1) any mental object is representable in a formal language; and (2) mental processes can find equivalents expressed dynamically through the execution of algorithms. This clarifies the extent and the intimacy of the possible co-operation, and suggests how this intimacy might motivate man to adopt, through a preoccupation with efficiency, a rationality according to which a given man's machine 'double' functions.

In forcing the trait of an emergent reality, an outline such as this sins, obviously, in having a degree of over-simplification. In the reality of research projects in progress, hypotheses that are more precise express, for example, psychological, linguistic and biological phenomena, with their respective formal models[13]. It remains nevertheless, whatever the restrictions that must be applied to that which we have discussed, that one does not see beyond the cognitive hypothesis what explanatory principle could give rise to a so an astonishing synergy, established before our very eyes, between man and the computer. It will be necessary in any case to monitor attentively the new work of these new interdisciplines, in order to see taking shape, perhaps in a precise way, the theoretical boundary between our mental (or cerebral) activity, and this part, or this reduction of itself, of which we

13 In a movement that tends to circumscribe, in bringing new objective elements to the philosophical debate, the immense enigma of the materiality of the mind.

would be capable of giving a formal description[14] and which we could simulate.

The question, which is eminently philosophical, relates to nothing less than the hypothetical limit of our capacity to build a representation of ourselves in the language of the automata. Its inevitable corollary is another question, which is not without social impact: in what way would man be irreplaceable? In other words, what would be the inexpungible position that he would occupy in association with the machine[15]? It is of this epoch that some questions raised in the philosophy of mind can have an economic relevance.

2.2 Socialisation of computational reason: Outline for a strategy of analysis.

That which precedes – cognition and computation being defined conceptually within the same formal framework – provides the theoretical argument that would explain the anthropological *possibility* of a universal recourse to data processing, and to its operational extensions, as a new technology. A necessary condition, to some extent, for this possible universality, but by no means a sufficient one. Because even if mankind had finally conceived for himself a tool that was ideally insuperable, the up-till-now irreversible character of its proliferation would still depend on the obviousness of factors of a wholly different nature: specifically, that singular or collective agents have the necessary knowledge, and that they decide to implement it by recourse to these new instruments; that they have the full capacity and means to implement it; and that they do it with a sufficient degree of success to

14 Art constitutes a field – an extreme one – that is open to these examinations. It is necessary in this regard to mention some research in progress on the cognitive analysis of the artistic phenomena. Cf. M. Borillo, J.-P. Goulette (Eds.), *Cognition et Création, Explorations cognitives des processus de conception*, Mardaga, Brussels, 2002.

15 The question of knowing *why* man is irreplaceable is of a whole other nature – ethical, psychological, economic or social, for example – but the answer to the scientific question of knowing *how* he is irreplaceable could well have an impact on the way we might respond to the first question.

be able to repeat their actions and to reinforce the evolution in progress by inciting imitation. This schema certainly does not account for the 'impact' of this new kind of technology, nor indeed for its 'risks' (a concept whose polarity it would be necessary to underline in recounting it to a subject, because that which threatens the interests of Peter may perhaps consolidate those of Paul – whilst not forgetting that 'all' is also a subject).

Nevertheless, in defining the dynamic of the proliferation through the *rôle* of its protagonists and through a set of conditions – be they economic, social, political, cultural, psychological or whatever – one of the problems posed by the evolution and the impact of cognitive technology is thus carried to where it must be, at the intersection, as yet little explored, between philosophy, the social sciences and the science of cognition. This displacement gives rise to the normal play of questions, by introducing a perspective that makes it possible to circumscribe the state of things with greater precision. For example, what are the categories of actors who can enter this process, and with what conditions? Consequently, who are those who are excluded from it? In which streams of information (of knowledge) is it necessary to bathe, in order to be able to claim to play such and such a *rôle*? Why are the crucial decisions made? (That is to say, according to what kinds of objectives – financial, political, social, technical, psychological, ethical or whatever – are they made?) How are they made; with what possible instruments? What positions of power engage them?

Other questions bring more definitely into play, in their internal logic, the nature of cognitive technologies themselves. For example: by what means is the transmission of knowledge established, which, beginning with the results of fundamental research, stimulates the technological derivations of those results? (Perhaps through the education system, but we must not forget the degree of military or industrial secrecy that prevails.) Some interesting feedback: in the cognitive domain, new expertise can affect the modes of diffusion and acquisition of knowledge. It is not only the technique that changes, but also the training of the men who work and live with it.

More than in any other field, knowledge is a major investment (or bet), and 'useful' research has long been subordinate to politics and economics: who fixes the thematic priorities, and according to what strategies? If we are interested in the inception and the diffusion of these technologies, we observe immediately that there are some branches or spheres of activity whose nature renders them particularly favourable for extremely variable but always very important reasons – for example, the omnipresence of symbolic forms of information (when databanks are already established), the existence of classical or original mathematical models that explain, for instance, the permeability of the money markets, of the management of stocks, of calculations for nuclear reactors, of weather forecasting, and so on.

In a revealing manner, the recent breakthrough in the mastering of the man-machine interaction (the importance of which will never be stressed enough) is certainly related to advances in the hardware, the explosion in the storage capacity and power of calculation of microprocessors, the so-called nanotechnologies. But it was made possible only by the merging in the software of a new knowledge coming from cognitive ergonomics, linguistics, intermodal communication, and so on, although they are far from their conclusion. One sees every interest in establishing priorities, making projections, foreseeing investment expenses, etc., of an analysis that crosses the specificity of domains with characteristics that define in depth the evolution of technology. But the social play is fundamentally polemical. The same movement will be necessary for anticipating and preparing to fight against the dismissals, and for demanding a share of the extra profits resulting from the reduction and the reorganisation of human labour – in short, to have some chance of successfully taking part in the *next* social debate.

We can already foresee that these technologies of a new type, because of their nature, will accompany us wherever we are and whatever we do. To this view without nuances, the immediate future will of course impose some restrictions, some limitations, and this 'company' will adopt, as suggested at the beginning of this text, some extremely varied forms and operational methods. We shall conclude there an already troublesome exercise; but in the face of such an upheaval, one does not see how to preserve the ratios of force which have allowed the play of history up to now, without a general awakening of the new nature of that which is presented fallaciously as a new technology, an additional technology.

Bibliography

M. Borillo, *Informatique pour les Sciences de l'Homme; Limites de la formalisation du Raisonnement*, Mardaga, Brussels, 1986.

M. Borillo, J.-P. Goulette (eds.), *Cognition et Création, Explorations cognitives des processus de conception*, Mardaga, Brussels, 2002.

N. Chomsky, *Le langage et la pensée*, Payot, Paris 1972.

J. Dubucs, F. Lepage (eds.), *Méthodes logiques pour les sciences cognitives*, Editions Hermès, Paris 1995.

P. Engel, Introduction à la philosophie de l'esprit, La Découverte, Paris 1994.

D. Laurier, *Introduction à la philosophie du langage*, Mardaga, Liège 1993.

J. McClelland, D. Rumelhart, *Parallel Distributed processing: Explorations in the Microstructure of Cognition*, MIT Press, Cambridge (Mass) 1986.

F. Nef, *Logic and language*, Editions Hermès, Paris 1988.

B.H. Partee, A. ter Meulen, R.E. Wall, *Mathematical Methods in Linguistics, Kluwer Academic Pub.*, New York 1990.

Mario Borillo, Institut de Recherche en Informatique de Toulouse, Université Paul Sabatier, Toulouse Cedex

Alessandro Bruschi

Logical models and the logical method

Abstract:

This article aims at defining the nature of logical models and the syntax with which they are built. Logic, regarded as mere philosophical tool – useful, at most, for clarifying formal or epistemological issues – has distanced itself from empirical research, which has chosen statistics and numerical analysis. Yet most problems in the social universe are not manageable experimentally or statistically, owing to the incapacity to physically alter the events, and to the lack of sufficiently large data distributions. When investigating current or past events, historians and sociologists are confronted with areas of inquiry characterised by one or few cases and many dichotomous variables. In such areas, they work simply by reasoning in a natural and implicit way.

The non-usage of an explicit logical method has produced serious consequences. On the one hand, indeed, we employ a logical methodology in a naïve way, without cultivating and improving on it. On the other hand, we are caught in the false dilemma of choosing between statistics and intuition, 'quantity and quality'. Giving up statistics should not be considered a free pass for a type of inquiry that is more literary than it is scientifically valuable, but rather the beginning of a different path as rigorous as the other. However, the method is extremely poor with respect to its power to provide information and confirmation. In particular, logical models – because they include variables with only two states and few cases – suffer from underdetermination, a lack of control over intervening variables, and a lack of empirical evidence.

Keywords:
method, logic, cause, analysis, model, test

1 Logic and social research

The term *model* has multiple definitions, yet they all share the meaning of a precise, although not perfect, but only partial or provisional, representation of an empirical phenomenon. The selected definition makes reference to the syntactic dimension of the representation; therefore, by a model of a social phenomenon, I intend the system of relationships that expresses the syntactic skeleton of the studied phenomenon. It is understood that a representation (and a model) is also something else: it has an empirical and theoretical content.

The model's syntactic structure involves the choice of a mathematical language. The mathematical language that I shall consider is that of logic. I define a model as *logical* when it consists of dichotomous variables and relationships interpreted by means of functions belonging to a logical calculus[1], as a model is *numerical* when it is composed of interval variables and relationships interpreted by means of functions belonging to a numerical calculus.

A model is produced and justified, as is any other scientific proposition. The sum of the procedures and rules with which logical models are rationally built is what I call the logical method. Such a method is characterized by the use of logic. However, I do not intend to suggest that other methods, such as the experimental, are illogical, or that their operations cannot be interpreted in accordance with a logical argument. I call a method logical when it employs, in its operations, only logic for its description, analysis and testing of models, in contrast to other methods, which use it only in a complementary way, or as an interpretation of scientific activities. More specifically, logic is used as a tool both of representation of the events in the model (modelling) and of its construction (model building). In the first case, it is used to describe the relationships between variables in the world (logical rather than numerical); in the second, it comes to affirm them. The two aspects are closely connected.

The language of logic is that which is most used and least cultivated in social research. The work of the sociologist consists, as does

1 Logical variables can also be polytomous. Nevertheless, this test is restricted to dichotomous variables, excluding polyvalent logic.

that of the historian, in the construction of logical models that are logically sustained and described. The paradox is that while 80% of research is neither statistical nor experimental, yet nevertheless potentially logical, little or no care is given to the use of the language of logic. In order to understand the nature of, and the reasons for, its lack of success, I shall put forward some considerations on the method.

2 The method and its characteristics

A research project is made up of various activities: observation, analysis, control, linguistic formulation, and so on. Different criteria can be used to classify it, such as the nature of the activities it employs, or the rôle played by some of these activities. I shall consider a research project's general justificatory force as a basic criterion of classification (Bruschi, 1998, Ch. 5). Such force depends on the degree of control that the researcher has over the factors that influence the results of the investigation.

Control is *manipulative* when it applies directly to the events of the universe: on the one hand, we physically eliminate with it any undesirable events, any disturbances; and on the other hand, we produce and modify those of direct interest for the research: we change the state of specific variables in the experiment, excluding or including them physically (as the physicist does in his laboratory). Control is *selective* when the researcher is limited to choosing which cases to study (as the sociologist does in the sample investigations). Selective and manipulative controls are *material*, because they are practised directly on the universe.

Control is *linguistic* when it is exercised on the information extracted from the universe; not so much on its events as on their symbolic representation. At the base of operations of statistical or logical analysis, control virtually eliminates the influence that some variables have on others, or it examines the coherence between the actual states of the variables and those they would have if they belonged to certain relationships.

These different control abilities characterize research methods, which are, fundamentally, reduced to two: experimental and systematic[2]. The first applies to manipulative material control; the second is limited to selective material control, or it applies to only linguistic control. While experimental investigation is typical of the natural sciences, systematic investigation is typical of the social sciences.

2.1 The experimental method

In the natural sciences, the reference model of investigation is the experiment, the Galilean method based on 'experience and reason' (Galilei, 1638). It seeks to physically modify the states of the variables that are the object of interest, and to exclude all other variables. In laboratory experiments, the researcher attains the maximum level of control.

Even if the complete isolation of a situation is an ideal that we can only ever approximate (as we can always suppose that there are unknown or uncontrollable influences), in the natural sciences we can neutralize most of the variables that may have some effect on the relationship being studied. The remaining disturbances have only minimal effects on the results of the experiment, displacing them slightly up or down in relation to the Gaussian distribution curve. The *ceteris paribus* clause has a high degree of plausibility.

The reproducibility of the phenomenon facilitates the repeatability of the experiment in a practically unlimited number of tests. The activity of analysis is readily connected to that of observation: we observe and analyse simultaneously. This gives the researcher great freedom to create specific conditions for testing a hypothesis. The modifications concern both the state of the independent experimental variables, and the introduction or exclusion of other variables, in order to investigate thoroughly the nature of a given relationship.

2 E. Nagel refers to 'controlled investigation', which has the same general meaning as 'systematic research' (Nagel, 1968, pp. 461-467: translation and citation from Italian version).

The repetition of the experiment in similar and different conditions makes experimental reasoning highly convincing, because its evidence is strengthened in accordance with quantitative (enumerative) and qualitative (eliminative) criteria. The researcher is able to organize the experiment in such a way that the experimental cases are characterized (at least in any relevant manner) by only the intentionally selected properties. If we introduce only *p* and *q*, and we find that *p* variations are accompanied by *q* variations, then the variations of *q* are imputable only to *p*, and not to some other variable (because there is none). Therefore, the experimental method enjoys the maximum internal and external validity – internal because the effects are imputable only to the causes; and external because the results are generalizable to all analogous situations (which facilitates the formulation of laws).

The success of this method has fascinated researchers from disciplines other than the natural ones. Social scientists have tried to introduce it into their sphere of activity (Campbell and Stanley, 1966; Cook and Campbell, 1979), but in the absence of the very reasons for its success: manipulability of events, isolation of variables, repetition of experiments, and variation of the circumstances. The properties being studied are not extractable from the personalities of individuals, and the so-called control group (by the nature and number of its components) has scientific sense in the biological sphere, but a great deal less in the social one. Also, repeatability is compromised, because it applies to different cases, or to the same cases *but* modified by previous experience of the experiment, or by changes in the environment. All this invalidates or substantially weakens the performance of the experimental method. For this reason, it would be more appropriate to speak of a *para-experiment*.

2.2 The systematic method

When the method is systematic, the checking of hypotheses is fulfilled through the observation of the universe *as it is* or *as it has been*, in contrast to the experimental method, which effectively creates a suitable universe to be observed. The cases are *given*, scattered in the worlds of past and present, and the variables are packaged, as it were, in the cases. Control is not *in corpore vili*; it becomes, above all, linguistic, exercised on the symbolic representation of the realized events. The argumentative strength of the investigation decreases.

The systematic method is divided into two main categories: statistical and logical. Each one is characterized by specific activities of analysis, control, and observation.

2.2.1 The Statistical method

The methodology of the statistical scientist is clearly distinct from that of the experimental scientist. His operations apply to cases belonging to *historical contexts*. Contrary to what happens with laboratory experiments, here we are not able to create specific conditions for testing a hypothesis. Materially, we can only select, by means of sampling, the number and quality of cases required, intentionally including only those variables of interest for the research; but the cases, besides being characterized by the explicitly considered variables, are also characterized by other, implicit, variables, either unknown, or known but disregarded. The relationships observed are not between p and q, but among p, q and a set S of n other variables. The *ceteris paribus* clause has to be replaced by that of *ad includendum:* we cannot rid ourselves of the context.

The linguistic control follows the survey. We don't have here a 'material reasoning'; we have only a linguistic one, exercisable on those variables that have been included in the survey plan. In this research there is a break between the activity of observation and that of analysis: they are separate and in succession; the fracture allows the distinction between *field* research and *desk* research, which become two distinct enterprises. The interaction between researcher and universe no longer follows an interactive process 'question – answer – question, etc.', but is fulfilled only through new research on a new sample, at times belonging to a different context. This is the reason for the great importance of planning in social research. The consideration of the variables to include in the survey scheme, and the quantity and quality of the necessary cases for analysis, must be the object of extreme care. The success of the investigation depends on what we have planned to observe; errors in planning are remediable only with difficulty.

A loss of control results in a lesser degree of internal and external validity. In fact, the coefficients of dependence that we find among variables in surveys are notoriously low. Samples and universes are *idiographic*; the cases are extracted not from the Nomological Universe of the laboratory, but from profoundly divergent historical situations. The repetition of the experiment, in accordance with the motto 'test and retest' of the Accademia del Cimento[3], turns into a complex data-analysis. It follows that statistical reasoning is decidedly less convincing than the experimental reasoning of the natural sciences.

2.2.2 The Logical method

The applicability of the statistical method requires the availability of a sufficient number of homogeneous cases; otherwise its operations become meaningless. Unfortunately, the majority of social research does not satisfy this condition: the historian and the sociologist, particularly when they leave the current epoch, find more and more research regions characterized by few cases (and many variables). They operate on these cases by way of natural reasoning, partly tacit, which is, potentially, the seed of what I have called the logical method.

Abandoning statistics, we enter into the vast and ambiguous region of the qualitative, whose meaning is sometimes non-scientific, with consequences both at the practical level of research and at the theoretical level of the method. At the practical level, we employ procedures and techniques that are neither executed with great care, nor improved. The attention of the researchers has been focused on the statistical and experimental methods, leaving the logical method to common use. Its proximity to natural knowledge and its continuous use, even in everyday life, have made it banal. It follows that the more

3 The academy was founded in Florence by the grand duke of Tuscany, Ferdinand II, and was the first academy devoted specifically to the experimental method (1657).

usual kind of research is the less cultivated methodologically. It is just because reasoning is a natural endowment that we know it so little.

In the meantime, we fall into the false dilemma of the choice between quantity and quality, statistics and 'interpretation'. The logical method allows us to escape from this dilemma. The abandonment of statistics must not to be considered a free pass, as it were, for investigations whose merit is often more literary than scientific, but rather the beginning of a different, equally rigorous, path. If research is founded on common sense alone, its processes belong to cognitive psychology, and not to the logic of research.

The choice between the statistical method and the logical method is determined by the number of cases, and not by the quality of the variables[4]. In systematic investigations we use the logical method when we do not have a sufficiently large body of homogeneous data for use with the statistical method. The applicability conditions for the logical method are, then, extremely generous, given that it requires few cases, and is satisfied with dichotomous variables. However, if the choice is decided by the number of data elements, what characterizes it is the syntax, the logic, applied in the formulation and building of the model. Of course – as I have already suggested – even when we operate experimentally or statistically, we reason. The fact is that here we *only* reason, and it is this that characterizes logical investigation.

Passing from the statistical method to the logical, the force of scientific argumentation decreases still further. Such a method is the most rigorous response that we can give to many research problems for which an experiment or a statistical survey cannot be employed, but it is extremely poor with regard to its justifying and informative efficacy. The defects, already recognized in statistical research, are enhanced here: the cases, being few, present a limited qualitative variety; therefore, the control process of the relationships is unable to take into account all of the circumstances that it would be opportune to consider.

The three methods are not in opposition, but in co-operation: the experimental method also uses the statistical and logical methods,

4 Dichotomous variables are used in statistics too, when they are sufficient in number. Their number is what is important, and not their type of scale.

while the statistical method uses the logical method. Any research project has many aspects and is made up of a lot of different problems to be confronted with appropriate techniques. The methodological rationality is *instrumental*: the choice of the means must be made solely with a view to optimizing the cognitive objective in the specific circumstances in which the researcher operates.

3 A historical perspective

The reasons for the paradox of a research project that unknowingly uses the logical method without paying it sufficient attention are to be sought, other than in the natural use of logic, in its history (Blanché, 1970; Kneale and Kneale, 1962), and in the history of the natural sciences, and also in the relationships that the social sciences have had with both of them.

Logic, born, broadly speaking, in the fourth century, was developed as a science of reasoning, independently of mathematics. It reached its apogee with Scholasticism: the dialectic, part of a *trivium* with grammar and rhetoric, was the principal instrument of knowledge. Considering that knowledge was revealed, above all, through the interpretation and analysis of texts, logic became the cornerstone of the cognitive enterprise. True science deals with universals, noumena, primary qualities; phenomena relate to secondary qualities. All that belongs to the terrestrial world, which springs from sensations, empirical phenomena and their observed processes, was not, continuing in this respect the Greek tradition, true science.

The rejection of logic occurred with the development of empirical knowledge. The conclusions of scientific argumentation are founded on observation and experiment, and no longer on deductions from basic principles. Moreover, in the words of Galileo Galilei, 'the great book of nature is written in mathematical characters' (Galilei, 1623). The representation of the natural is realized through the relationships between variables, and requires mathematical tools. In the modern age, logic therefore loses its centrality in favour of observation and mathematics (and of rhetoric for men of letters). The new research method is based on *'sensed experiences'* and *'certain demonstrations'*. The operations of experiment obscure the logical processes on which they are founded. Science *uses* logic, but does not *cultivate* it. It is individuated in Scholasticism, in theology, in the past: it is the tool of philosophy, as mathematics is the tool of the new science.

The long sleep of logic was to end only in the second half of the 19th century, with its formalization (Boole, 1847; Frege, 1879)[5]; but it held a different position in the panorama of knowledge. Up until then, logic had been the interest of philosophers, and mathematics the interest of scientists; now logic concerns the mathematicians. Its new field of application was the systematization of mathematics, and the aim was to try to reduce all mathematics to logic. So logic became *mathematical logic.*

During the same period, besides the 'logicization' of mathematics, we had the 'mathematization' of logic with Boole's algebra. Its mathematization opened a new field of application. Until then, logic had been only a demonstrative tool. Now it became a syntactic one too, suited to describing phenomena through logical functions linking dichotomous variables. Logic acquired the function of modelling.

In the modern age, if deductive logic lost much of its importance, inductive logic, generally neglected in the classical and medieval ages[6], acquired a new rôle with the experimental sciences. Bacon innovated inductive logic (Bacon, 1620), introducing the eliminative criterion, which would be used by J.S. Mill and then by von Wright[7], in addition to the enumerative criterion. Bacon's innovation regarded inductive logic not only as an inferential science, useful for interpreting the research activity of scientists (as deductive logic already was), but also as a direct tool for empirical investigation, the principal research

5 In the modern age, Leibnitz made an important contribution by to the symbolization of logic, but his attempt had no followers.

6 In spite of the fact that in the mediæval age philosophical attention focused on deductive logic, there was the important work of Roger Bacon and Robert Grossateste, who influenced Francis Bacon (Crombie, 1953) and, through him, J.S. Mill.

7 While the enumerative criterion produces positive evidence to affirm hypotheses, the eliminative criterion seeks to falsify them. A true hypothesis is just one that we have not been successful in proving false.

method of the scientists, thus continuing the mediæval tradition; yet as an investigative method for the physical world, it is unquestionably conservative, even if it remains relevant in the justificative processes, or rational reconstructions, of the operations of the scientists.

At the end of the nineteenth century, following the development of probability theory, inductive logic was formalized too, becoming probabilistic. As such, it improved its inferential power and acquired a new area of application in statistics. In that way it co-operated meaningfully in the development of what we call the statistical research method, which joins the experimental one. It operates in the field of great numbers, collectives, the distributions of data that form matrices in which statistical calculation may be practised. The scientist operates through sampling, inferring the results of data-analysis.

Following this historical path, we are able to understand the reasons for the refusal to use logic as a research instrument for the social sciences. In the antipositivist tradition, logic is not rejected for the love of irrationality; it is accepted in its natural version as part of human reasoning, but is rejected in its artificial, formalized version. The antipositivist tradition uses logic as a normal process of argument, implicitly and in a natural way. Its connection to mathematics, after Boole and Frege, makes it extraneous to the 'science of the spirit'. It also becomes, in a certain sense, part of the 'quantitative', if only for the 'ideology' it expresses. *L'ésprit de finesse* wins over *l'ésprit geometrique*. In the myth of the literate sociologist – at times a committed intellectual, interested in problems of substance and meaning – there is no place for mathematical logic, and little for statistical inference.

In the positivist tradition of the social sciences, the ideal model of reference for research is that of the natural sciences, which use experiment and numerical calculation. The social researcher tries to approach them through so-called experimental designs (in the laboratory, in the field, and natural), but above all through statistical investigations. Logic belongs to philosophy; it is a useful tool with which to clarify formal or epistemological matters, extraneous to research. In particular, with logical positivism, formalized logic acquired a great rôle in epistemological reasoning for the interpretation and analysis of research operations, but it remained a metalinguistic function; research operations were still experimental, statistical and mathematical (in a traditional sense). With the crisis of neopositivism and the abandonment of the strong conception of science, the popularity of logic decreased in favour of rhetoric.

4 The logical method and its subdivisions

In the history of logic, and of knowledge in general, we can discover antecedents of the logical method. In Scholasticism, logic was *the method* of obtaining knowledge, but its employment occurred outside the constitutional framework of what we now call scientific knowledge. Bacon reformulated the logical method, applying it this time to the empirical world – but inappropriately, because the universe of the natural sciences requires above all the experimental method. Baconian logic, then, acquired an epistemological but not a methodological value. Even if the Royal Society considered Bacon among its tutelary deities, its own members, such as Newton and Boyle, used the experimental method.

The logical method finds its application consistent with the principle of maximum investigative efficacy only in the social sciences, in the field of systematic research, whenever the statistical method is inapplicable. Its explicit and deliberate use does not represent a methodological revolution, as did the introduction of the experimental and statistical methods, but rather an improvement and a strengthening of a technology transferred and reformulated from the 'natural' level of cognitive psychology to the 'artificial' level of scientific methodology.

The logical method enjoys abundant and diverse resources. These consist in classical logic, its modal developments, inductive logic (probabilistic or otherwise), informal logic, and many others, such as fuzzy or linear logic. However, these resources, be they deductive or inductive, may be adapted to the specificity of social problems. Given the extreme broadness of its application and the heterogeneity of its content (idiographic or nomological, macro or micro), the logical method assumes many different characteristics (Agodi, 1997). In very general terms (yet insufficient to classify specific applications), it can be developed according to a *strong* version and a *weak* version. The first is *formalized*, employing deductive and inductive logic expressly for the formulation and the testing of assertions. The second is *non-formalized*, employing natural language and informal reasoning, expressed in grammatical forms.

The non-formalized, weak form of the logical method, applicable to data and text alike, will not be dealt with here. I shall examine only a part of the strong version, staying within the limits of classical deductive and inductive logic. These two types of logic are used for different purposes. In the representation of the world through models, we employ only *deductive* logic. To confer evidence on the representation, to justify it, we use mainly *inductive* logic. More precisely, the following text concerns deterministic models, in which hypotheses are tested according to eliminative criteria.

The so-called eliminative methods take origin from the work of Bacon. Reformulated by J.S. Mill (Mill, 1872), and also called *canons* (of agreement, of difference, of residues, of concomitant variations, etc.), they were a considerable step forward from Bacon's work, even if Herschel had anticipated many of Mill's ideas (Herschel, 1830). Mill's methods abandoned every scholastic form and introduced the language of causes and effects, but they were limited by the use of a logic expressed in natural language. The limitations were overcome by von Wright in the mid-twentieth century (Wright, 1951), although his work had some precedence in Broad's theory of conditions (Broad, 1930). Von Wright expressed, in a strictly logical manner, the analysis and verification of assertions; he concentrated on the first three of Mill's canons, expressing them in terms of necessary and sufficient causes.

It is interesting to note how the philosophical tradition of causality was strong enough to prevail over the algorithmic approach; the literature of logical models is above all causal, speaking not about relationships among dependent and independent variables, but about causal relationships. I shall follow this use, which facilitates the understanding of the subject matter.

The great chapters of the logical method concern bivariate and multivariate analysis, and the reformulation of causal language. Multivariate analysis is used for the testing of bivariate relationships through the introduction of further variables, for the extension of the analysis to compound causes, and for the control of multivariate structures (i.e. trivariate and beyond).

5 The structure of the model

I define a model as causal when its variables are causally connected. A causal relationship is expressed through a conditional, and is thus characterized by implication, even if it is not simply reducible to it. A scientific relationship is something more of a mathematical function.

While I shall speak of 'model' with reference to language, I shall speak of 'system' with reference to the universe. Model and system do not always coincide. The universe is generally more complex than the language that aims to simplify it in order to analyse it in rational way.

I shall call a variable *connected* when it has a causal relationship with at least one other. By the *structure* of a model I intend the particular pattern of connections that characterize it. Structures are classifiable on the basis either of the number of variables and relationships involved, or of the configurations according to which the variables are connected (Bruschi, 1993, pp. 147-149).

Given n variables, the number of possible structures (ST_n) is given by the formula:

$$ST_n = 2^{n\,(n-1)}$$

With two variables, ST_n is equal to 4 (two monorelational configurations in opposite directions, one birelational, and one disconnected). With three variables, ST_n is equal to 64. This shows a very rapid growth in complexity as n increases. The high number of possible structures obliges us to make a definitive choice of model-structure to examine (and confirms the conviction that research should be theoretically oriented), because the possible structures of a system are too numerous to be investigated logically. The choice *excludes* some relationships from the analysis, because they are considered irrelevant or impossible), while it *includes* others either *positively* (they must be present) or *negatively* (they must be absent). Therefore, the testing of the model is fulfilled by the affirmation and negation of the relationships that constitute the hypothesized structure. In such a way, we reduce the set of possibilities to a subset that is more easily manageable; but as a consequence, the conclusions of the operations are conditioned, so to speak, by the assumptions made.

The logical method pays particular attention to the language of causality, investigating, first of all, the nature of the necessity and sufficiency of the causal relationships within the chosen model. In the numerical calculus we have dependent and independent variables, and functions which link their variations: there is no trace of this causal aspect, which modellistic logic instead examines. As we know, a cause is sufficient for an effect when its presence is sufficient to obtain the effect; so we can define a sufficient cause (C_S) as the antecedent of a conditional that has the effect as its consequent[8]. A cause is necessary, when it is not possible to get the effect without its presence; therefore, a necessary cause (C_N) is definable as the consequent of a conditional that has for its antecedent the effect. If, to the presence of a cause, there corresponds only one given effect, and to that effect only that one cause, we have a necessary and sufficient cause (C_{NS}), defined by the double implication:

$$C_S \rightarrow E$$
$$C_N \leftarrow E$$
$$C_{NS} \leftrightarrow E$$

These properties are fundamental for the control of the causal relationship, which is based on the presence and absence of the states of its variables.

Normally, we consider causes as *agent* causes – all causal theory follows the perspective of 'production' – but we are able to consider even *counteragent* causes. If the first causes *produce*, the second ones *destroy*: in this last case I shall speak of *inverse causes*. So, a sufficient cause is inverse when its presence provokes the absence of the

8 For the truth table of implication, see Table 1.

effect: $(c \rightarrow \neg e)$[9], and a necessary cause is inverse when it is not possible to get the absence of the effect without its presence: $(c \leftarrow\neg\; e)$[10]. We have a necessary and sufficient cause when, in the presence of the cause, there is an absence of the effect, and also in the absence of the effect, the presence of the cause: $(c \leftrightarrow \neg e)$[11].

In all the relationships previously considered, the presence of the antecedent caused the presence (direct implication) or absence (inverse implication) of the consequent. However, we can consider cases in which the *absence* of the antecedent causes either the presence or the absence of the consequent. I shall call these last causes *negative causes* to distinguish them from the preceding ones, which I shall call *positive*. If the absence of an event provokes the presence of another, the first negatively causes the second, and the negative implication is direct: $(\neg c \rightarrow e)$. It is inverse when the absence of an event produces the absence of another: $(\neg c \rightarrow \neg e)$.

Vague concepts such as *cause* and *effect* can receive a more exact scientific definition; consequently, their declarations are subjectable to more accurate methods of corroboration and confutation. The thought remains 'weak'; but objectively, not subjectively, because it is the strongest that we are able to employ in these research situations.

The logical functions used in models differ according to the definition or functioning of their structure. This distinction is present even in statistical analysis, in which the measures of dependence are correctly distinguished from the functions of dependence. The problem is not to find the combinations that falsify or corroborate the hypothesis, but rather to find the correspondence between the value of the independent variable and that of the dependent one, as happens in mathe-

9 In other words, the two events are contrary; they cannot be both present, although absent, simultaneously. In fact, this implication is equivalent to the contrariety $(c \,\underline{\vee}\, e)$, that is to the negation of the conjunction, $\neg(c \wedge e)$. I use the symbol '$\underline{\vee}$' to indicate contrariety, and '$\underline{\underline{\vee}}$' to indicate exclusive disjunction. For the truth tables of conjunction and inclusive disjunction, see footnotes 12 and 14.

10 The expression is equivalent to the inclusive disjunction $(c \vee e)$, which excludes only their contemporary absence.

11 The two events are incompatible: if one is present, the other has to be absent; they cannot be both present or both absent. The expression is equivalent to the exclusive disjunction $(c \,\underline{\underline{\vee}}\, e)$, that is to the negation of equivalence $\neg(c \leftrightarrow e)$.

matical algorithms. Here, the existence of the relationship is not in question, but only how one variable varies as a function of the other. In such a case, simple implication is not usable, because the calculation needs univocal correspondence. As is well known, the truth table for implication accepts three combinations: 1, 1; 0, 1; 0, 0 (Table 1). Consequently, either we use the double implication or we establish that the dependent variable maintains the preceding value when the antecedent assumes value 0 (considering positive implication).

Table 1

	P	*Q*	*P* → *Q*
I	1	1	1
II	1	0	0
III	0	1	1
IV	0	0	1

6 The testing of the model

Identifying the causal relationship syntactically with a conditional, it is possible to use the truth table of implication to examine whether there are instances with the combination that deny the relationship: when the antecedent is present and the consequent absent (Row II of Table 1). If an instance shows such associations of values, the causal relationship does not exist.

A bivariate model is testable with Table 1. With three variables *(P, Q, R)*, we use a trivariate truth table, which will display eight 'combinations', in accordance with the formula $2^3 = 8$, as in Table 2 below. Technically, they are dispositions with repetitions ($D'_{n,k}$). Thus, if k is the number of elements considered (for us the two truth-values, 1 and 0), and n is the number of variables considered:

$$D'_{n,k} = k^n$$

Therefore, to check the structure of a trivariate model of the following type (drawn from the possible 64 alternatives)

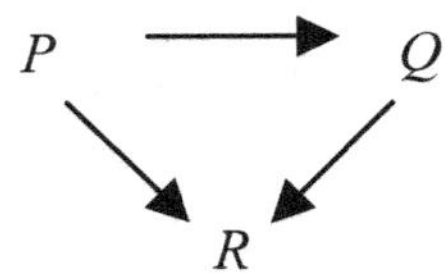

we have the following table, in which, for simplicity, *'PQ'* replaces *'P →Q'*:

Table 2

	P	*Q*	*R*	*PQ*	*QR*	*PR*
I	1	1	1	1	1	1
II	1	1	0	1	0	0
III	1	0	1	0	1	1
IV	0	1	1	1	1	1
V	1	0	0	0	1	0
VI	0	1	0	1	0	1
VII	0	0	1	1	1	1
VIII	0	0	0	1	1	1

The value 0 placed under the variables *(P, Q, R)* or the relationships *(PQ, QR, PR)* declares their absence; the value 1 their presence. These values are placed under the relationships in the same row as the values of the variables that affirm or deny them. For instance, the value 0 in column *PQ* and row III denies the existence of the *PQ* relationship. In fact, to deny an implication *PQ*, represented as ¬*PQ* in our notation, where '$\neg$' indicates 'not', we must have *P* present and *Q* absent: $P \wedge \neg Q$[12]. In such a way, we can hypothesize structures and check them via the selected cases that show the associations of values denying or affirming the relationship.

In Table 2 above, there are two truth-value combinations that deny each relationship, owing to the fact that every value-combination of two variables conjoins with a third variable whose value can be either 0 or 1. For instance, $\neg PQ$ is present in row III, in which *R* (the third value) has value 1, and in row V, in which *R* has value 0. Generally, the number of combinations falsifying a relationship (F_n), where *n* is the number of variables, is given by the formula:

$$F_n = 2^{(n-2)}$$

Thus, with two variables, we have only one combination ($F_2 = 2^{(2-2)} = 2^0 = 1$), the second of the bivariate table (Table 1). With three variables, we have two combinations, as the trivariate table (Table 2) shows. With four, we have four combinations, and so on. This is easily comprehensible if we realize that the introduction of each new variable doubles the number of combinations (the succession is geometric).

The analysed case is that of a positive structure, in which the considered relationships must all be present, but we can also have a negative structure, whose relationships must all be absent, and a mixed structure, in which some relationships must be present and others absent. For this last structure, testing is rather more complicated (Bruschi, 1993, Ch. 6).

12 A compound sentence $P \wedge Q$ (conjunction) is true only when *P* and *Q* are both true. The symbol '$\wedge$' corresponds to the word 'and':

	P	*Q*	$P \wedge Q$
I	1	1	1
II	1	0	0
III	0	1	0
IV	0	0	0

7 The bivariate analysis

The testing of causal relationships is possible in two basic ways: one founded on the causes (causal correspondences C), the other on the effects (effective correspondences E). With C, you select the cases according to the presence or absence of the supposed cause; with E, you select them according to the presence or absence of the effect of the supposed cause. The *modus operandi* of C is typical of the natural scientist, who manipulates the hypothesized causes to observe their effects. That of E is typical of historical research, in which events are first attested, and then have their causes sought.

In each case, the researcher examines the correspondence between the values of the antecedent and those of the consequent of the conditional in which the causal relationship consists, in such a way that is always possible to get the second combination of the truth table for implication (Table 1); otherwise we eliminate the possibility of rejecting the hypothesis. In these operations we always follow the APCA principle (Antecedent Present, Consequent Absent). In fact, if we select cases by antecedent, we use the first two combinations of Table 1 (the antecedent is present while the consequent is present or absent); if we select them by consequent, we use the second and the fourth combinations of Table 1 (the consequent is absent, while the antecedent is present or absent).

7.1 Sufficient causes

To test for a sufficient cause (e.g. Is the proportional electoral system a sufficient condition for political instability?) we select cases in which the supposed cause is present, and then see if the effect is also present, in accordance with the criterion of *sufficient reason*, C (Table 3). Otherwise, we select cases in which the effect is absent, then see whether the cause is absent too, in accordance with the criterion of *inverse agreement*, E (Table 4).

Table 3

SAMPLE	CANDIDATE PROPERTY	CONDITIONED PROPERTIES			
Cases	C	E_1	E_2	E_4	E_n
1	1	1	0	1	1
2	1	1	0	1	1
3	1	1	0	1	0
4	1	1	1	0	0
.	1	1	.	.	.
m	1	1	1	0	1

Criterion of sufficient reason

Table 4

SAMPLE	CANDIDATE PROPERTIES				CONDITIONED PROPERTY
Cases	C_1	C_2	C_3	C_n	E
1	1	0	0	0	0
2	0	0	0	0	0
3	1	1	0	0	0
4	0	0	0	1	0
.	.	.	0	.	0
m	0	1	0	0	0

Criterion of inverse agreement

In the above tables, we have selected m cases in order to explore the relationships. In the first table, we have considered many effects with the same possible cause; in the second, several possible causes with the same effect. In Table 3, C proves a sufficient cause of E_1; in Table 4, C_3 is a sufficient cause of E. The other relationships have been eliminated.

7.2 Necessary causes

To test for a necessary cause (e.g. Is the market economy a necessary condition for political democracy?) we select cases in which the effect is present, to see whether the cause is present too, in accordance with the criterion of *direct agreement*, E (Table 5); otherwise, we select cases in which the cause is absent, to see whether the effect is absent the too, in accordance with the criterion of *necessary reason*, C (Table 6). Always following an exploratory orientation, we have selected *m* cases in both the tables. From the inspection we infer that in Table 5, C_2 proves a necessary cause of *E*; in Table 6, *C* is a necessary cause of E_3.

Table 5

SAMPLE	CANDIDATE PROPERTIES				CONDITIONED PROPERTY
Cases	C_1	C_2	C_3	C_n	E
1	1	1	0	0	1
2	0	1	1	0	1
3	0	1	0	1	1
4	1	1	1	1	1
.	.	1	.	.	1
m	1	1	0	1	1

Criterion of direct agreement

Table 6

SAMPLE	CANDIDATE PROPERTY	CONDITIONED PROPERTIES			
Cases	C	E_1	E_2	E_3	E_n
1	0	0	0	0	1
2	0	0	0	0	0
3	0	1	1	0	0
4	0	0	0	0	0
.	0	.	.	0	.
m	0	0	1	0	1

Criterion of the necessary reason

Which criterion of the two classes to use is purely a matter of pragmatics; it depends on the availability of the cases and on the ease of the procedures. The criteria are distinguished only by the different value-combinations of the variables used to test the relationships, always in accordance with the APCA principle.

8 The multivariate analysis

Although the bivariate analysis remains fundamental in clarifying the nature of the causal relationship, social phenomenology is not correctly analysable by using only bivariate tools. Social investigation does not rely on the laboratory, where the relationship is undisturbed, and its universe presents an integrated systems of variables. Therefore, it is always wise to doubt results based on bivariate analysis alone, and to check them with the aid of multivariate analysis.

8.1 The testing of bivariate relationships

Relationships that hold at a bivariate level can prove spurious when a third variable is introduced. For example, even if fever and cold are strongly correlated, the fever is only apparently caused by the cold; the true cause is a virus that simultaneously provokes fever and cold. Suppose, for example, we have tested at a bivariate level that *P* implies *Q*. In the table of observations there is no case of *P* without *Q*. Nevertheless, the observations show, in the selected cases, that a third variable, *R*, is present, giving rise to the hypothesis that *R* is the true cause of *Q*. Consequently, we are able to go on to test a new structure with three variables. Searching further cases in which *R* is absent fulfils the test: if ever *Q* is absent when *P* is present, the *PQ* relationship is spurious.

We may also have doubts whether the confirmed relationship between *P* and *Q* is mediated, not direct. The direct relationship between two properties is frequently a simplification, because between the two there exists a causal path to be made explicit. For instance, I may suspect that the relationship between poverty (*P*) and juvenile crime (*Q*) is not direct, but mediated by the intervening variable family break-up (*R*). Unfortunately, this time the testing is less satisfactory than in the preceding case. If the path between *P* and *Q* is mediated by *R*, the structure is described as a chain '$(P \rightarrow R) \wedge (R \rightarrow Q)$'. This configuration is inconsistent with the negation of '$P \rightarrow Q$' (affirming the existence of only the mediate path), because the model would become contradictory. By the transitive law of implication, if '$P \rightarrow R$' and '$R \rightarrow Q$', then '$P \rightarrow Q$'.

8.2 Compound causes and effects

We can eliminate a property as a cause *not* because it is extraneous to the effect, but because it is only *part* of the cause; it is, we may say, a *partial cause*. If the cause is compound, the single component properties are not causes in themselves – singularly, they have no efficacy. It is their combination with other properties that constitutes the cause (not a simple property). In accordance with Mill, 'rarely, not to say never, does this invariable sequence subsist between a single consequent and a single antecedent. It subsists, usually, between a consequent and the sum of rather a lot of antecedents'[13].

The new truth table to check the compound cause is based on the conjunction of the simple properties for the sufficient one, and on their disjunction for the necessary one (Skyrms, 1966, Ch. 4), both in accordance with the CADIC principle (Conjoin the Antecedents, Disjoin the Consequents)[14].

The composition of a sufficient cause is based on conjunction, because the disjunction would be banal (when one of the disjoined components is already a sufficient condition, their disjunction is auto-

13 The statement is ambiguous because it is not clear whether it refers to a compound cause or to concomitant causes (Mill, 1888, Vol. II, p. 459: translation and citation are from Italian version).

14 A compound sentence $P \vee Q$ (disjunction) is true when at least one of its member is true. The symbol '$\vee$' corresponds to the word 'or':

	P	Q	$P \vee Q$
I	1	1	1
II	1	0	1
III	0	1	1
IV	0	0	0

matic)[15]. For instance, while an accelerated growth-rate (C_1) may not be sufficient to produce political instability (E), when combined with an elevated value of disproportion in the distribution of income (C_2) and the presence of organized political formations (C_3), the conjunction may be sufficient to produce E.

The table that follows is based on correspondences according to the criterion of inverse agreement, E. With it we affirm that in m instances the conjunction of C_1, C_2 and C_3 ($C_1 \wedge C_2 \wedge C_3$) holds as sufficient cause of E.

Table 7

SAMPLE	CANDIDATE PROPERTIES				CONDITIONED PROPERTY
Cases	C_1	C_2	C_3	$C_1 \wedge C_2 \wedge C_3$	E
1	0	1	1	0	0
2	1	0	1	0	0
3	0	1	0	0	0
4	1	0	0	0	0
.	.	.	.	.	0
m	1	0	1	0	0

Criterion of inverse agreement

The composition of a necessary cause is by disjunction, excluding the conjunction. The use of conjunction adds no further information to that which each component already provides singularly. If C_1 and C_2 are singularly necessary causes, so is their conjunction; therefore, it is sufficient to analyse the simple conditions. It is not banal to consider their disjunction, because C_1 can share some quality with C_2 such that each one of the two properties is a 'surrogate' of the other; in such a case, the presence of either of the two is sufficient (that is, their disjunction) as a necessary condition of E. For example, although the road infrastructure (C_1) is not a necessary condition for developing an area, it may be disjoined with other infrastructure types, for example, railways (C_2).

15 The expression 'disjunction', without a qualifying adjective, means 'inclusive disjunction'. See footnote 12.

Table 8 is based on correspondences in accordance with the criterion of direct agreement, E. By examining the table, we affirm that, although neither C_1 nor C_2 is by itself a necessary condition, their disjunction is.

Table 8

SAMPLE	CANDIDATE PROPERTIES			CONDITIONED PROPERTY
Cases	C_1	C_2	$C_1 \vee C_2$	E
1	1	0	1	1
2	0	1	1	1
3	0	1	1	1
4	0	1	1	1
.	.	.	1	1
m	1	1	1	1

Criterion of direct agreement

8.3 The Boolean tool

An approach, which has received widespread approval is Ragin's qualitative comparative analysis (QCA). The approach has the merit of being able to examine compound and multiple causes together (both positive and negative ones), using Boolean algebra (Ragin, 1987, 1989, 1993)[16].

Considering *n* properties and *m* cases, we select those instances in which the *n* properties (be they present or absent) correspond to the presence of the effect, and these are assumed to be a sufficient compound cause (Table 9)[17]. Consequently, the *n* properties (present or absent) in every one of them are connected by conjunction.

16 As we saw previously, a property can be a complete or partial cause; therefore, considering more properties we can have only a compound cause or more causes. Moreover, they may be causes when present or when absent.

17 The notation in the tables is not that used by Ragin. It has the advantage, however, of not requiring a specific introduction and explanation.

Table 9

SAMPLE	CANDIDATE CAUSAL PROPERTIES						CONDITIONED PROPERTY
Cases	C_1	C_2	C_3	C_4	C_5	C_6	E
1	1	1	1	0	1	1	1
2	1	0	1	0	0	1	1
3	1	0	1	0	0	1	1
4	1	0	0	0	1	1	1
.	.	.	.	.	1	1	1
m	1	0	1	0	1	1	1

In the second place, we 'sum' all these different compound causes by disjunction, therefore allowing each member its autonomous causal efficacy. Limiting our consideration to only two compound causes, cases 1 and 2 from Table 9, we have the following expression:

$$(C_1 \wedge C_2 \wedge C_3 \wedge \neg C_4 \wedge C_5 \wedge C_6) \vee (C_1 \wedge \neg C_2 \wedge C_3 \wedge \neg C_4 \wedge \neg C_5 \wedge C_6)$$

Subsequently, the method simplifies such an equation, eliminating the redundant properties of the compound causes by way of the Quine-McClusky algorithm (Grimaldi and Pauletto, 1996). If a component property of a compound cause is present or absent while the other properties are constant and the effect is present, that property is discarded as redundant. For instance, if we have the two following causes, each giving rise to the same effect:

$$\mathrm{C}_1 \wedge \neg \mathrm{C}_2 \wedge \mathrm{C}_3$$
$$\mathrm{C}_1 \wedge \mathrm{C}_2 \wedge \mathrm{C}_3$$

we can reduce them to the simpler expression $C_1 \wedge C_3$, as the presence or absence of C_2 makes no difference to the existence of the effect.

In such a way, we can come to define a minimal causal equation, both in the number of its members (causes) and their internal composition (number of properties). Through the associative property and De Morgan's law, we can extend our knowledge to equivalent causal configurations[18].

18 According to De Morgan's law, $\neg(p \wedge q)$ is equivalent to $(\neg p \vee \neg q)$.

This interesting research strategy, which follows an exploratory case-based approach, needs, however, to be located within the logical theory of conditionals; otherwise it risks losing its falsifying power, and also risks confusing sufficient causes with necessary ones. Moreover, it has no means with which to analyse the internal structure of a compound cause, and it excludes causal models that are more complicated than those based on a simple relationship between a compound cause and an effect.

9 Limits of the analysis

In many aspects, a logical model provides less information than a numerical one. Logical variables merely inform us of the presence or absence of a property; numerical ones inform us of many other states that the property can assume, between which the distances are known. Consequently, even the functions that connect numerical states are much richer and more numerous than those connecting logical ones.

In addition, when we have more than one sufficient cause of a given effect, the effect is overdetermined. The logical calculation has no power to resolve problems of *concausality*. Indeed, because the logical variables are dichotomous (with values 0 or 1), an isolated cause or many causes together may produce the same effect. If the members of a firing squad strike a convict's heart simultaneously, it is not possible to discriminate the effect of each bullet that struck him, in order to calculate 'partial coefficients'. To be killed by a single rifle or by a firing squad is perfectly equal.

Other limits arise for the transitive property of causal relationships, since implication, which syntactically constitutes it, is transitive. We cannot test a causal chain by cutting it. In a causal chain, every link, irrespective of its distance from the effect, has the same causal claim as any other. We are not able to stop and go backwards: people say that the industrial expansion of England was caused by its people's love for raisins, and that Cleopatra's nose determined the future of the Roman Empire (Hanson 58). When the variables are numerical, a regression coefficient affords sound reasons for stopping along the path;

but when they are dichotomous, all their intensities are flattened out to their presence or absence, and setting bounds becomes impossible.

Some social dynamics prove rather banal. Tendencies towards development or decadence are not describable: with only two values, we cannot model the growth or involution of a system. Even the behaviour of cybernetic systems, which defend themselves against environmental changes, has an impoverished representation. While in quantitative language systems, stabilizing processes are clearly manifest, logically, they are not appreciable. A stabilized system becomes one that maintains always the state '1', as with one that 'grows', or the state '0', as with one that 'decreases'.

But the great limit of causal logical analysis is due to the fact that, as the number of variables increases, the cases at our disposal become too few to satisfy the number of combinations necessary for the control of the relationships, as the comparativists know all too well; this weakens the empirical evidence of the conclusions. We are in the presence of a poor methodology, and we have to take this into account.

Bibliography

M.C. Agodi, 'La ricerca su base logica', in L. Ricolfi (ed.), *La ricerca qualitativa*, NIS, Roma 1997.

F. Bacon, *Novum Organum*, 1620, It. transl., F. Bacone, *Nuovo Organo*, in Id., *Opere filosofiche*, Laterza, Bari 1971.

R. Blanché, *La logica e la sua storia*, Ubaldini, Roma 1973 (1970).

G. Boole, *Mathematical Analysis of Logic*, 1847, It. transl., *Analisi matematica della logica*, Silva, Milano 1965.

R. Bradley, N. Swartz, *Possibile World*, Hackett Publishing Co., Indianapolis 1979.

C.D. Broad, 'The Principles of Demonstrative Induction', in *Mind*, 39, 1930.

A. Bruschi, *La competenza metodologica*, Carocci, Roma 1998.

A. Bruschi, *La metodologia povera*, NIS, Roma 1993.

A. Bruschi, *Metodologia delle scienze sociali*, Bruno Mondadori, Milano 1999.

D.T. Campbell, J.C. Stanley, *Experimental and Quasi-Experimental Designs for Research*, Rand McNally, Skokie (IL) 1966.

Th.D. Cook, D.T. Campbell, *Quasi-Experimentation*, Houghton Mifflin Company, Boston 1979.

I.M. Copi, C. Cohen, *Introduction to Logic*, 1994, It. transl., *Introduzione alla logica*, Il Mulino, Bologna 1997.

A.C. Crombie, *Robert Grosstesta and the Origins of Experimental Science*, Clarendon Press, Oxford 1953.

A. Fisher, *The Logic of Real Argument*, Cambridge University Press, Cambridge 1988.

J.B. Freeman, *Dialectics and the Macrostructure of Arguments*, Foris Publication, Berlin 1991.

G. Frege, 'Begriffsschrift, eine der arithmetischen nachgebildete Formelsprache des reinen Denkens', 1879, It. transl., 'Ideografia. Un linguaggio in formule del pensiero puro a imitazione di quello aritmetico', in Id., *Logica e aritmetica*, Boringhieri, Torino 1977.

G. Galilei, *Dialogo sopra i due massimi sistemi del mondo*, 1632, in Einaudi, Torino 1970.

G. Galilei, *Discorsi e dimostrazioni matematiche intorno a due nuove scienze*, 1638, in Boringhieri, Torino 1958.

G. Galilei, *Il Saggiatore*, 1623, in Feltrinelli, Milano 1979.

L. Griffin, C.C. Ragin, 'Some Observations on Formal Methods of Qualitative Analysis', in *Sociological Methods & Research*, XXIII, 1, 1994.

R. Grimaldi, M. Pauletto, 'Un metodo per la produzione di modelli logici: l'algoritmo di Quine-McClusky', in R. Grimaldi (ed.), *Tecniche di ricerca sociale e computer. Modelli, base di dati e basi di conoscenza*, Omega Edizioni, Torino 1996.

N.R. Hanson, *Patterns of Discovery*, Cambridge University Press, Cambridge, It. transl., *I modelli della scoperta scientifica*, Feltrinelli, Milano 1978.

J.F.W. Herschel, *A Preliminary Discourse on the Study of Natural Philosophy*, Longman, Rees, Orma, Brown & Green, and John Taylor, London 1830.

G.E. Hughes, M.J. Cresswell, *An Introduction to Modal Logic*, 1968, It. transl., *Introduzione alla logica modale*, Il Saggiatore, Milano 1973 (1968).

W.C. Kneale, M. Kneale, *The Development of Logic*, Clarendon Press, Oxford 1962, It. transl., *Storia della logica*, Einaudi, Torino 1972.

M. Marsonet, *Introduzione alle logiche polivalenti*, Abete, Roma 1976.

J.S. Mill, *A System of Logic Ratiocinative and Inductive*, J.W. Parker, London 1872, It. transl., *Sistema di logica*, Utet, Torino 1988.

M. Mondadori, M. D'Agostino, *Logica*, Bruno Mondadori, Milano 1997.

E. Nagel, *The Structure of Science*, Harcourt, Brace and World 1961, It. transl., *La struttura della scienza*, Feltrinelli, Milano 1968.

C.C. Ragin, *Fuzzy-Set Social Science*, University of Chicago Press, Chicago 2000.

C.C. Ragin, 'Introduction to Qualitative Comparative Analysis', in T. Janoski & A. Hicks (eds.), *The Comparative Political Economy of the Welfare State*, Cambridge University Press, New York 1993.

C.C. Ragin, *The Comparative Method. Moving Beyond Qualitative and Quantitative Strategies*, University of California Press, Berkeley 1987.

C.C. Ragin, 'The Logic of Comparative Method and the Algebra of Logic', in *Journal of Quantitative Anthropology*, I, 1989.

L. Ricolfi (ed.), *La ricerca qualitativa*, NIS, Roma 1997.

W.C. Salmon, 'Probabilistic Causality' in *Pacific Philosophical Quarterly*, LXI, 1890.

B. Skyrms, *Choice and Chance*, Dickenson Publishing Company, Belmont (Ca) 1966, It. transl., *Introduzione alla logica induttiva*, Il Mulino, Bologna 1974.

P. Suppes, *A Probabilistic Theory of Causality*, North Holland, Amsterdam 1970.

S.S. Thomas, *Practical Reasoning in Natural Language*, Prentice Hall, Englewood Cliffs 1986.

D.N. Walton, *Informal Logic*, Cambridge University Press, Cambridge 1989.

G.H. von Wright, *A Treatise on Induction and Probability*, Routledge and Kegan, London 1951.

Alessandro Bruschi, Dipartimento di Scienza della Politica e Sociologia, Università di Firenze

Luciano Celi

White bears and washing-machines: models (and theories) of concepts

To Raffi, because 'Die Philosophen sind oft wie kleine Kinder, die zuerst mit ihrem Bleistift beliebige Striche auf ein Papier kritzeln und dann den Erwachsenen fragen "was ist das?"'

Wittgenstein

Abstract:

What are concepts? In which way can we grasp them? We will discuss various conceptual models and theories that try to answer these questions in different ways. The approach is 'historical' because it highlights problems and benefits of each theory. Having a concept is often a way to say we are able to make correct categorization.

Similarity plays a fundamental role in this process that is involved in different ways: in Prototype Models it is only a mechanism to explain categorization, but in a recent model of concepts, the Hybrid Model, this role changes and explains in which way observing regularities means being able to open a space for nomological or theoretical aspects. Why does the colour of washing-machines and bears – assuming that it is the same colour: white – play a completely different role in our explanation? The Hybrid Model answers this (and other) question(s).

Keywords:
concepts, conceptual theories, categorization

1 Introduction

Theories and conceptual models, as other research fields, must satisfy a series of requirements, that need to be mentioned, in order to evaluate their ability or their effectiveness in explaining the phenomena concerning concepts.

This paper is an attempt to follow a path, which is inevitably arbitrary and somehow historical, aiming at showing the reasons why, in recent times, the idea of a Hybrid Model for concepts has arisen among some researchers.

A 'good' model should therefore be able, for instance, to explain how we can have concepts, how we can use them, how it is possible to define, as regards objects, categories of belonging, or to explain inferences we do every day. These few elements are sufficient to constitute valid criteria to test a theory, highlighting its limits and its qualities.

Clearly this paper only constitutes a partial analysis. As a matter of fact, the position of Connectionism about these issues is never mentioned.

2 The Classical Theory of concepts

The peculiarity of this 'historical' theory of concepts is the claim for them to be defined: in order to consider concepts as definite, they should meet some necessary and sufficient conditions that guarantee their applicability. To 'apply' a concept essentially means to indicate a psychological process through which an object falls under that concept. We will proceed step by step: the necessary and sufficient condition to have a coffee from a vending machine (possibly not out of order!) is to insert the correct amount of coins. In other words: until the correct amount of coins is not inserted into the machine (accomplishment of the necessary condition) it will be impossible to have coffee but as soon as coins are inserted, coffee will be served (accomplishment of the sufficient condition)[1]. For instance, we consider the concept of 'telephone': in accordance with the Classical Theory, we can think of it as a mental representation where some necessary and suffi-

1 This example comes from Boniolo, Vidali (1999), 507. In this case, we only have a necessary condition (to insert the correct amount of coins), that becomes a sufficient condition too. It is appropriate because it can be a metaphor of conceptual definition where the coffee/concept must be served/defined if and only if we introduce in the vending machine the correct amount of coins/properties. The lack of only one of them has as a consequence the lack of definition.

cient conditions are specified, in order to indicate all the objects that fall under this concept: so, 'telephone' must have certain characteristics such as being *equipped with a receiver for speaking and hearing, having the possibility of getting a connection with a telephone network* and so on.

These components specify conditions that the object should meet, so that we can call it a telephone (and possibly use it as a telephone): these are called necessary conditions. These features should then occur together, so that we can say that they are jointly sufficient for the definition of the concept: the absence of only one of them means, referring again to the case of the vending machine, not having inserted a sufficient amount of coins in the machine.

2.1 A Set Model for the Classical Theory

It is certainly true that not all concepts have simple definitions like 'telephone' or 'bachelor'; on the contrary, we can surely state that the majority of concepts we use is very complex. The strategy through which the Classical Theory of concepts can account for the conceptual complexity consists in the attempt to reduce complex concepts to more simple concepts, through a sort of decomposition of basic concepts: Katz states that the word

> 'chair' can be decomposed into a set of concepts which might be represented through semantic markers: OBJECT, PHYSICAL, NON-LIVING, ARTEFACT, FURNITURE, PORTABLE, SOMETHING WITH LEGS, SOMETHING WITH A BACK, SOMETHING WITH A SEAT, SEAT FOR ONE[2].

This definition, that can be formalized according to a Set Model, has two important consequences: the first one leads us to think that, in order to have the concept of 'chair', it is necessary to have an experience of all these components and that, to put an object into the chairs category, all these features must be met. In this way, concepts are clearly defined and somehow *Boolean*: either an object falls into a category

2 Katz (1972), 40.

because it accomplishes the required features for that category, or it does not belong to that category.

This aspect, that we can call *verification*, has a direction going from the object to the (theoretical) characteristics of a certain category; we will call *justification* the opposite direction which goes from the features of the category to the object. In other words, we can say that we are 'justified' to believe (to think, to know) that a certain object falls under a certain category if the features defining that category are satisfied by the object before us.

The second consequence, that can be seen as a kind of corollary of the first one, concerns the mechanism of conceptual acquisition: the decomposition made in the example suggested by Katz shows how, once the semantic markers for 'chair' are acquired, that specific concept has also been acquired *tout court*, because we are able to recognize and place that kind of objects in the real world. Semantic markers are in this sense selective and allow us to make the distinctions that constitute the basis of our ability to categorize. Incidentally, I would like to remark that most of the attractiveness and beauty of the model explaining the Classical Theory is based upon the 'economy' of its descriptive principles: the underlying mechanism, explaining the conceptual acquisition – the decomposition of meaning – is the same underlying the explanation of another important phenomenon of cognition: our inferential ability.

Consider now these three sentences:

(1) Giorgio is an unmarried man. So Giorgio is a man.
(2) Giorgio is a weigh-lifter. So Giorgio is a man.
(3) Giorgio is a bachelor. So Giorgio is a man.

Sentence (1), unlike sentence (2) is analytical, in the Kantian sense, and this guarantees the conclusion following from the introduction. Sentence (3) could produce some doubts: it is more similar to (2) than to (1) but, thanks to the possibility of decompose its meaning, we are able to determine the constituent elements of the concept 'bachelor' into the concepts 'man' (or into the further decomposition 'male', 'adult') and 'unmarried'. This allows us to understand whether our inferences are valid or not, and therefore to understand how many of the inferences we do every day are incorrect – usually many more than we can think.

The fulfilment of necessary and sufficient conditions, together with our inferential ability, leads to the *semantic possibility* of concepts:

> when one acquires the concept ROBIN, doing so crucially involves acquiring a concept that *refers* to robins. And when one draws an inference from ROBIN to IS A BIRD, or IS AN ANIMAL, one draws an inference *about* robins[3].

This possibility of the semantic reference shows, first of all, that the necessary and sufficient conditions play an 'operative role' in the model: an object (the robin) can fall under a certain concept (the bird) if certain conditions and priorities are fulfilled. The recognition of the properties defining the object guarantees the belonging to that concept and, in this case, the object acquires its 'meaning'. A meaning that guarantees, as we have seen, a series of inferential features: to have a robin in front of us means to already know that we are in front of a bird and an animal.

2.2 Problems of the Classical Theory and its Model

The most famous objection to the definition of concepts, according to the model proposed by the Classical Theory, is brought by Wittgenstein[4]; he does not define the concept of 'game' according to the rigid constraint of the necessary and sufficient conditions anymore but, after having highlighted the different kinds of game and the different features that the necessary (but not sufficient) conditions can offer as a possible definition, is inclined to a different conclusion: games constitute a sort of 'network' that links, with no hierarchy, the different items: each example of game has some features in common with the others, a sort of 'familiarity'. The family resemblances then become the metaphor used by Wittgenstein in order to show that it is possible to trace some common features among the members belonging to the

3 Margolis, Laurence (1999), 13.
4 Wittgenstein (1953), §§ 65-67.

family, even if every single element that composes it is different from the others.

In other words, a concept like 'game', with so many different features that change, arise and disappear, cannot be defined in accordance with the criteria that the Classical Theory would use: we actually know the concept of 'game' without knowing a definition in terms of necessary and sufficient conditions; this leads to think that concepts, to be considered as such, do not necessarily refer to definitions.

Another problem is due to conceptual typicality: this topic has sapped the foundations of the Classical Theory more than any other and has paved the way for the coming of new theories, born in the 70s from the report of a series of experimental results. What emerges from experimentation is the typicality effect: Rosch and Mervis (1975) have carried out a research which started from subjects with difficulties in classifying objects according to their *goodness* or typicality as members of a category. The notion of typicality is rather intuitive and consists in the measure of how much – expressed either with a percentage or with a figure – an object represents the category to which it usually belongs. A classical example that is usually found in the literature regards, for instance, the typicality of 'apples' belonging to the 'fruit' category. If we can reasonably consider apples as the most typical fruit of the category (to which we can attribute 100% typicality), we cannot do the same for 'olives' and 'tomatoes'.

The experiment of Rosch and Mervis (1975) consisted in submitting to subjects lists of properties of elements from different categories; some properties occurred in many lists, while others occurred less frequently. The experiment's result was a measurement of the foreseen typicality (as an hypothesis) from the distribution of properties that occurred in those lists: an instance is considered the most typical (of its extension) if its properties are common to the category it belongs to.

The problem that these experimental results brought to the Classical Theory is the lack of natural models accounting for this effect. If we want to represent graphically the most typical elements of a category – for instance the birds category – in terms of centrality in the set, then the robin will stand in the centre of this set, while the penguin will stand in a peripheral area, being a less typical bird than the first one. If we attempt an analogous representation for the Classical Theory, the space location of robins and penguins is completely insignificant within the set because both are birds and both have predictively fulfilled the necessary and sufficient conditions in order to belong to that set. This inability – that comes from the constraint of the necessary and sufficient condition to determine the belonging – causes a loss of information because we are interested in models that are able to take into account how people think about categories and we should not exclude that many people do not classify the penguin as a bird, or the tomato as a fruit or the whale as a mammal.

3 The Prototype Theory

In the 70s, the focus of the scientific community shifted from the Classical Theory – unable to explain the typicality of events – to a theory called the Prototype Theory. Although it has a considerable number of versions – resulting in different models – this theory claims that concepts are complex representations, whose structure encodes a statistical analysis of the properties the members (belonging to those concepts) tend to have. The possibility of using the statistical instrument allows a higher degree of freedom, in comparison with the constraints put by the Classical Theory: it is possible that the examples representing a concept do not have the characteristics which are more specific to the concept to which those examples belong to:

> For instance, if BIRD is composed of such features as like FLIES, SINGS, NESTS IN TREES, LAYS EGGS, and so on, then on the Prototype Theory, robins are in the extension of BIRD because they tend to have all of the corresponding properties: robins fly, they lay eggs, etc. However, BIRD also applies to ostriches because even though ostriches do not have all of these properties, they have enough of them[5].

So, if the Classical Theory needed to produce necessary and sufficient conditions, in order to understand whether an element falls under a certain concept, the Prototype Theory simply needs a sufficient

5 Margolis, Laurence (1999), 27-28.

amount of properties, where some of them will obviously be more significant than others.

The Prototype Theory, unlike the Classical Theory, represents an advantage as regards important explanatory aspects because it slackens some constraints that the latter puts to conceptual features. As a matter of fact, whereas the Classical Theory is not able to provide us with any solution for the lack of the definitions that many concepts have, the Prototype Theory does not requires that concepts have a definitional structure and this allows to avoid (and maybe explains and foresees) the difficulties that the theorists have had in trying to set definitions. According to the Classical Theory, as we have seen, analyticity constitutes an important point but in recent times it has undergone an important revision that has put its importance under discussion; as a consequence, while this revision directly involves all explanatory elements that the Classical Theory found on analyticity as well, it does not touch at all the Prototype Theory, because it plays no particular role in it. This aspect brings with it another advantage for what concerns non-demonstrative inferences: a function of concepts is to allow to people to take important information in order to categorize correctly an element or an instance. As a matter of fact, on the basis of what we have said about the equivalence of topological spaces within a set, for the Classical Theory a concept must encode a certain number of information – that will all have the same 'value' – on its elements; however having no selection instrument to understand *which* features are the most typical to be encoded, it should include such a great number of them that a concept becomes impossible to use. The solution found by the Prototype Theory is that concepts must encode the distribution of *statistically relevant properties* in one category; since the criteria are statistical, (1) such properties are not put on the same hierarchical level and (2) only those that are really important to establish the belonging of an object to a category are selected. By selecting according to statistically relevant properties, concepts having a prototype structure are able to generate many more inferences than the classical representations can.

3.1 A Model for the Prototype Theory

As anticipated in the introduction, the formalisation through models constitutes the attempt to provide for something that we can analyse, starting from theoretical assumptions: so, models, like theories, have their own value if they are able to give adequate explanations and predictions on aspects which are important for cognition. The following model – called Accumulator Model – stems from the formalisation of the notion of similarity, through the notion of distance[6] and verifies the similarity of a certain object with respect to a target category. From an operational point of view, some comparisons between features sets shared both by the object and by the category are being carried out. Once ascertained that the characteristics of the examined element belong to the reference category, the positive and/or negative values are 'accumulated', on the basis of the presence or not of the examined characteristic.

> The idea behind this principle is that the estimated similarity of any two items, *i* and *j*, is measured by comparing the sets of shared and distinctive features that are associated with them. Where I and J are the feature sets, the function can be defined as follows:
>
> $$\text{Sim}(I,J) = a f(I \cap J) - b f(I - J) - c f(J - I)$$
>
> The constants a, b, and c allow for different weights to be assigned to the set of common features (I∩J) and to each set of distinctive features (I–J and J–I), and the function *f* allows for weights to be assigned to individual feature. To illustrate how the principle works, consider the measure of similarity between BIRD and TWEETIE, where the latter is a representation that, for simplicity, incorporates just four features: FLIES, SINGS, IS SMALL, and LAYS EGGS[7].

6 Some criticism regards also the distance as a way of formalizing similarity: the concept of 'comet' is – in a conceptual space – closer to the astronomic concept of 'star' or to the concept of 'Christmas'?

7 Margolis, Laurence (1999), 30.

In order to show more simply how the model[8] works and how the calculation of the distance of elements from the target category is done, consider the following example:

- The sets of common and distinctive characteristics have the same weight (=1) – consequently, the constants of the general formula a, b and c all equal 1;
- The function *f* assigns to each features the same weight;
- We have the following table:

Feature	Bird	Robin	Chicken	Vulture
Flies	Yes	yes	No	Yes
Sings	Yes	yes	No	No
Lays eggs	Yes	yes	yes	No
Is small	Yes	yes	No	No
Nests in trees	Yes	yes	No	Yes
Eats insects	Yes	yes	No	No

Applying the similarity formula we can calculate the distance between two exemplars – robins and vultures – or make the comparison between an exemplar and a superordinate category[9]. If we consider the similarity between 'bird' and 'robin', for instance, we will get: 6–0–0=6, whereas between 'bird' and 'chicken' the result is 1–5–0=–4.

The simplified model proposed here highlights important characteristics: distance is a mechanism subjected to similarity and similarity

8 We use this Model for his simplicity. Other researchers (see, for example, Cohen and Murphy 1984) give a Prototype Theory formalization with other models that have more complex rules and the application is less immediate.

9 Margolis, Laurence (1999) highlights that 'in the literature, *exemplar* is used to denote subordinate concepts or categories, whereas *instance* in used to denote individual members of a given category' (footnote 30, page 25).

is a mechanism subjected to classification, so the more two objects are similar, the less distant they are. This simple rule can explain both the typicality and the non-typicality of an object: if we know that BIRD 1 (B1) is a very typical bird and that BIRD 2 (B2) is a little distant from it, we can infer that B2 is similar to B1. If B3 is a little distant from B2 – and consequently from B1 – we can infer that also B3 is sufficiently typical. In a similar formalization we can, for instance, also ascertain the value of distance |D| (indicated as absolute value) to establish in which neighbourhood we could consider typical elements or less typical elements and so on, so that we can indicate precisely what we mean by *distant*, *little distant*, etc. If we know that B4 has a |D| value which is higher than the one that separates B1 from B3, then we know that it is a less typical bird. If B5 is very near to B4 – and consequently far from B1–B3 – it will be a less typical bird, and so on.

The construction of this topological space takes place through a process of accumulation, and the measurements for the distance seems somehow reliable.

A similar mechanism can be carried out by comparing lists of properties instead of real exemplars: if B1, the most typical bird, has a series of properties constituting 100% of the typicality for that category, the more B2, B3 etc. have properties similar to those of B1, the more typical they are.

Distance also predicts another aspect on which experimental psychologists are particularly keen: the rapidity of classification. There seems to be a law of direct proportionality between this rapidity and distance: the shorter the distance between elements, the more rapid will the response of classification be. In the same way, in less clearer cases – as for instance for the fading categories in the example of Labov (1973) – the computational workload increases because the number of comparisons (between the object observed and the target category/ies) is superior for the determination of belonging or not to a certain category. If the object looks like a finger cup, but also has the typical features of a glass (if, for instance it is tall and narrow) this object is, from the point of view of the topological representation, equidistant from the most typical cup and from the most typical glass.

3.2 Problems of the Prototype Model

There are problems that the Prototype Model is not able to solve. We know that many concepts have an incomplete or generally unknown definition; if we take the concept of 'measles', the properties that people usually associate to this disease as symptoms – high temperature, skin rash etc. – are, in general, reliable effects and, as such, they can thus be decoded for a prototype representation. At the same time, since the symptoms do not constitute a disease but rather are the effects of a variety of causal interaction, they cannot be considered as a totally unreliable guide for the presence of the disease. Someone could show the symptoms of a disease, without actually having the disease, and someone else could have the disease without presenting its symptoms. A bird with all its feathers torn off is still a bird and a tiger with three paws, domesticated, with no teeth and whitish still remains a tiger, whereas it is not convincing that a toy-tiger is a tiger.

In other words, the representation by prototypes does not determine the correct extension of a concept like 'bird' or 'tiger'; on the one hand, it is not sufficiently wide to include all the birds and all the tigers – because of atypical elements that can be considered very different from the typical examples of the category – so different that are not included, as it could happen in the case of the whitish tiger. On the other hand, it is somehow too rich, so rich as to incorporate information including things which are not birds or tigers – as in the case of the toy-tiger. Another harsh problem is that we have notions and concepts that do not even have a prototype, as, for instance, 'the present king of France', 'a sax quartet of IV century' or 'Cindy Crawford's great-great-great grandchild'. These are all things we can imagine, for which there could be no prototype at all.

3.2.1 The fuzzy logic and the problem of compositionality

A model which takes inspiration from the Accumulator Model is constituted by the formalisation through sets that follow to the fuzzy logics' rules. We can create an extensional model in which, for instance, the concept of bird is represented through a quadruple <B, TYP_{BIRD}

(X), b, d> consisting in a B class of birds, a typical TYP_{BIRD} function measuring the degrees of 'birdness' creating maps of all the birds in the B in the continuous range [0,1] a prototypical bird b, and a metric distance d between the members of B. Here, a concept is mainly represented through the assignment of a typical value TYP_{BIRD} *(X)* to each *x* object. Although the central typicality factors are well specified, this tells us little about how the concept 'bird' is mentally represented. The TYP_{BIRD} function is given in his extension, that is there are no specific rules capable to actually calculate a TYP_{BIRD} *(X)* for each *x*. There is no syntax or lexical information in the model – not even where the name 'bird' appears – and nothing is said on the typical properties of birds, as, for instance, having wings, flying, or concerning the relations between *bird* and other concepts, for example the superordinate category 'animal'. Although the quadruple <B, TYP_{BIRD} *(X)*,b, d> includes the b prototype, none of the rules concerning the concepts formation actually uses b, and does not even depend on the knowledge domain. In other words, this separation of concepts from the knowledge domain, specific to the extensional approach, *disconnects* empirical data from their formalisation.

A concept, in a hypothetical extensional model of this kind, is not a *representation* of a category or of a set, but is the set itself: this approach does not take into account real cognitive processes, as we all agree with the fact that people who recognize birds, do not have the whole set of birds 'stored' in their memory; in this sense the cognitive aspect recalls closer the metonymical models of Lakoff (1987). As Cohen and Murphy (1984) claim, it still remains obscure how, giving a merely extensional description of classes of objects in a certain domain can explain the way in which people conceive these objects, or how it can represent concepts originating from the classification of those objects.

Moreover, this model meets serious problems when the concepts to be represented are not simple but have several constituents. In fuzzy extensional models, concepts like A–and–B are composed of the conjunction of sets associated with A and B. In order to form conjunctive concepts the fuzzy logic suggests us to use a rule (the so-called *minimum rule*) that gives typicality to the x concept for which, according to the formalism previously introduced, $TYPp(x)$ is a characteristic function measuring the typicality of element x – that we are examining – in relation to the P set. So, if the typicality of a concept is somehow determined by the typicality of its constituents, we can apply the minimum rule to the concept x: it says that the degree of typicality is given by the intersection of the degree of typicality of its constituents, that we can call A and B. We can express this typicality as $(x)(TYP_{A \cap B}(x))$. In this regard, the belonging degree of the x element to the concept 'red apple' is determined by:

$$TYP_{RED \cap APPLE}(x) = \min [TYP_{RED}(x), TYP_{APPLE}(x)]$$

So, if an object has its characteristic value of 0,9 for 'apple' and 0,8 for 'red', its value for 'red apple' will be the minimum value between those two values: 0,8. In order to form the complex concepts 'red apple' or 'pet fish', the minimum rule does not require a knowledge of prototypes or of exemplars composing the categories. For this reason the knowledge for which *apples* are *fruits* and that *fruits* have a *colour*, that *red* is a *colour*, or that *red fish* is an example of *pet fish,* is not required.

An apparently innocuous implication of the minimum rule lays in the fact that the typicality of an object is always lower than the typicality of the single characteristics examined:

$$TYP_{A \cap B}(x) \leq TYP_{A}(x), TYP_{B}(x)$$

This inequality is also highlighted by the Classical Set Theory, in which the characteristic function is reduced to the binary values 0 and 1. In this way, it is assured that if something is not a good A, it will surely not be a good A–B. Osherson and Smith highlight that this implication is false for the categories of the real world: for example, red fish is more a typical pet animal than it simply is a fish or a pet animal and this invalidates the minimum rule. The minimum rule is clearly inadequate: the aspect that arises more problems with regards to this model concerns the compositionality, and the formalisation through the models we have described seriously suffers from this limit.

Moreover, the compositionality surely is an important feature of the conceptual system and provides for the best explanation for a fundamental aspect of human thinking: its productivity.

4 The Theory-Theory

The most characterizing aspect of this theory of concepts concerns knowledge, that is generally assimilated to the scientific reasoning. This aspect has at least two important qualitative consequences: first, the categorization deviates from the most recent empirical models consisting of processes to verify an example by a list of sensory properties. Second, the analogy with the scientific reasoning highlights the affinity between concepts and theoretic terms; so the discussion about the latter ones can be borrowed by psychology. I think it is necessary to explain at once what we mean by using the word theory, in order to maintain a clear distinction between the domains of science and psychology, or as an extension of cognitive sciences.

The Theory-Theory was born as an explanation and as an alternative, new, theoretic support to the Piagetian theories of acquisition and conceptual development in children, thus it refers to the general meaning of theory, which has necessarily to be distinguished from the specific meaning of scientific theory. The most general meaning involves, indeed, using an often cited expression, a 'web of beliefs' that is not bound to a specific domain, it is transversal to the domain of knowledge, not necessarily consistent in itself, and it is able to explain perceptive aspects as well. All these elements are not suitable for a theory in the various scientific domains. In spite of these differences, the results achieved by researchers to explain the acquisition and the conceptual development in children seem to be able to fit, with the above mentioned limitations, the conceptual variations that take place in science: some elements of acquisition seem to be of value apart from the context we insert it in.

Susan Carey, whose research is mainly about understanding processes in children and the differences as regards the learning in adults, says that: 'Concepts are the constituents of beliefs; that is, propositions are represented by structures of concepts. Theories are complex mental structures consisting of a mentally represented domain of phenomena and explanatory principles that account for them'[10]. Thus, if maybe we are not yet able to say what concepts really are, no doubt we are able to say what they are not. For sure concepts are not lists of elements bound to necessary and sufficient conditions, as the Classical Theory maintained, and neither are they mere probabilistic distributions of features or properties, as the Theory of Prototype and related models maintained. Concepts mostly concern things in the world and they show structured relations between them; the relations must be explained not only by observing their frequency or distribution, but also by weighing up the causal aspect.

Consistent models of belief – and inconsistent ones as well[11] – are crucial to understand the nature of concepts and the way they develop. As a consequence of the work carried out in the sixties about the acquisition of artificial concepts, we inveterately tend to talk about concepts in terms of entities consisting of combinations of features or which are the result of the summary of processes involving specimen or/and dimensions. Although the most recent classical-probabilistic views refuses any constraint to the necessary and sufficient conditions to define natural concepts, it is true that the prepositions such views are based upon, consist of maintaining, more or less silently, that the meaning of concepts can be fully described by lists of features and probabilistic processes, or correlations between these features (or exemplar or dimensions).

Many researchers highlighted the probabilistic approach limitations in the interaction with conceptual combinations (Osherson and Smith, 1981); whereas the combined concept structure is often difficult to foretell only by the probabilistic structure of its elements according to the compositionality aspect. Moreover, many phenomena bound to conceptual variation in children cannot be easily explained in terms of probabilistic models (as widely discussed in Keil 1987 and Carey 1985). In previous views this theoretic inadequacy is evident and stirred an interest in the relation between concepts and theories in psy-

10 Carey (1991), 258.

11 See Frixione (1994), in particular chap 10 (235ff)

chology, also because of the important role these two elements always played in the Philosophy of Science.

According to Quine (1969, 1973) human beings are the only ones endowed with a particular ability that enables us to go gradually beyond our inborn 'animal sense of similarity', and, thanks to memory, to reorganize similarity spaces according to significant relations from a theoretic point of view. From this point of view, concepts are in the first place probabilistic classifications of the signals produced by sensory transducers and they are mainly associationistic (hence the importance of similarity); they are inserted in theories, when time and competence allows it.

When theory reaches its explanatory limitation, alternative representation models are necessary: in many aspects this process is similar to what happens in science. For example, just think about the theory of universal gravitation in physics. It explains many micro/macrocosm phenomena but it is inadequate in the case of some dimensions – subatomic particles – so that an alternative explanation model is required: the quantum mechanics. It is important to notice that in science these models often include previous theories as well. We can give many examples: the first one I remember is about geometry; those who knows some trigonometry knows at the same time that Pythagoras' Theorem is a particular case of Carnot's Theorem of triangles in general.

One of the symptoms of the inadequacy of a scientific theory is the arising of some *ad hoc* postulates (for example think about the postulates that astronomers used for centuries – the issue of epicycles to explain the apparent movement of some planets, such as Venus, that followed retrograde trajectories in the vault of Heaven, in order to justify observed celestial phenomena before the coming of the heliocentric theory, or about the arbitrary introduction of the cosmologic constant Λ by Einstein, in opposition to the idea of an expanding universe), whereas the new theory has a causal explanatory knowledge that is self-sufficient and uniform and often supported by a sort of formal elegance, at least in scientific theories. Assuming that in our representations of objects elements with associative contents[12] can coexist with

12 Think about how many theorems in geometry are based on similarity.

elements having more theoretic contents, we need to better understand the way these two methods interact.

Problems concerning the nature of intuitive theories and how they are represented, have to be related to systems of beliefs as well: for example in children many phenomena seem not to be considered but at the same time we can say they have some theories. Keil (1989) gives the example of many children attending nursery: they would answer that they have no idea of the way you catch a cold, or food makes us grow up, or why many children resemble their parents; and if you ask them to choose between two contrary alternatives explaining the possible mechanisms of these phenomena, they sometimes prefer clearly incorrect mechanisms. Once I went with the son of some friends of mine to a heliport: a helicopter was taking off and he naturally asked me, by obvious analogy with the battery-operated toy he played with at home, where the batteries were and how many of them were necessary for such a big helicopter. In this case the child never doubted that the working mechanism could be different; he simply extended his theory about the working mechanism of his helicopter to a superficially similar object.

Concepts are surely inserted in the web of beliefs but not as static nodes; they are relational, dynamic and maybe also explanatory because they provide for an interpretation of the associative matrix they refer to. Discussing these phenomena, researchers divided them into three groups:

- *Conceptual restructuring in a strong sense:* the cosmologic revolution by Copernicus and Galileo is the most cited case; the apparent movements of planets such as Jupiter and Venus are wholly reconceptualized in the new (heliocentric) theory, with no link with the previous (geocentric) theory. On this subject Kuhn (1962) discusses the incommensurability between these two concepts referring to the same phenomenon. Nevertheless there is an opposite process, called *coalescence,* by which two phenomena kept apart in a previous theory gather. For example remember when Galileo put together *violent movement* and *natural movement* that were considered as separate elements in the Aristotelian Physics;
- *Conceptual restructuring in a weak sense:* thanks to the competence acquisition in a domain, (for example playing chess), new relations between concepts and new schemes substitute the old ones, in order to enable the solution of new problems and variation of the solutions of the old ones.

Playing chess we notice a great difference between beginners and experts: both of them knows the rules of the game but the expert has a really different view (schemes, concepts, inferences) as compared to the beginner. The conceptual restoration in a weak sense often reflects this beginner/expert distinction, although it is necessary to have precise criteria to understand if we have to do with this sort of restoration.

- *Simple reconceptualization:* simple properties are sometimes reanalysed in terms of other relations, like when Newton reconceptualized the weight of a body as a force exerted by the Earth on an object. Another example referring to the 'force' idea in physics is provided by Hertz in '*"Prinzipien der Mechanik"* where he suggested not to seek a direct answer to the question "what is force?" but to find a new settlement of the Newtonian physics avoiding the use of the force concept as fundamental.'[13]

4.1 Beginners and experts: weak meaning of the conceptual change

Assuming that the above mentioned formation models are correct, we should be able to think about knowledge development as a single process starting during childhood and continuing in science. According to Theory-Theory upholders, the term 'science' is best understood as a psychological phenomenon in continuity with other phenomena, rather than as an abstract, logic system by which we look for the truth.

We know that one of the main differences between adults and children is that the latter ones have limited knowledge of the world and can be considered as 'beginners' in many domains in which adults are 'expert'. One of the main effort of cognitive science was the attempt to understand the beginner/expert change in adults, and some particular domains, such as chess competence, physical sciences (especially mechanics) and the solution of mathematical problems, were analysed.

The research by Chi, Glaser and Rees (1982) mainly focuses on the analysis of perceived similarities between the elements in a domain; beginners and experts were asked to group physics problems according to their typology. The results showed that the two groups faced the problem in a very different way; on the one hand, beginners focused on the superficial features of the problems, subdividing them according to the kind of involved objects (inclined planes, pulleys, etc); on

13 Monk (1990), 440.

the other hand experts grouped the problems according to their possible solution, distinguishing problems that could be solved thanks to Newton's rules, energy conservation rules, etc. These two classifications are in a certain way orthogonal because problems about, for example, pulleys can appeal to different physical rules but, beyond this, it seems that experts organize their knowledge in a deep, structured way that seems impossible for the beginners.

These differences, however, shouldn't lead us to think that two concepts of the same categories are incommensurable (in the sense suggested by Kuhn, 1962); concepts such as *force, energy*, etc. can be identified both by beginners and by experts and must be easily 'translated' from one system to the other, otherwise, from a theoretical point of view, we would not be able to explain phenomena such as misunderstandings, wrong ideas, etc. (which normally occur in learning); only thanks to this 'compatibility' between two conceptual systems we are able to explain how a beginner becomes an expert.

If it is like this, we are able to:

- conceive beginners' misunderstandings and wrong ideas as beliefs of the same physical dimensions that are different from the experts' ones, or how beginners represent relations of the same physical dimensions in a different way as compared to experts;
- endowed with sympathy beginners, when they direct their own choices to correct equations to apply to in a given problem, although they aren't fully competent;
- think about a prioritarian restoration at a basic level – a subordinate change level; this description assumes that these concepts, which are at a basic level for beginners and subordinate for experts, must be the same in both systems.

5 The Hybrid Model

Similarity, conceived as an explanatory construction concerning concepts acquisition, recently played a vital role, also thanks to the Theory-Theory. Its dimension was reconsidered thanks to the recent discussion upon a model with an eminent predecessor[14] called Hybrid

14 See Quine (1969).

Model. This model inherits a set of experimental observations discussed in the debate on the Theory-Theory and comes out as a model capable to explain and distinguish the most important aspects of knowledge: the acquisition and use of concepts. The name itself suggests its dual vocation; indeed, two aspects have been highlighted by researchers: the theoretic element responsible for rules acquisition and their structural coherence and the atheoretical element merely bound to the perceptive aspect. According to these considerations related to the Theory-Theory it seems that concepts consist of a mix of both explanatory systems and atheoretical property classifications.

The limitations of explanatory systems or theories in organizing properties and concepts are better understood with a concrete example. Keil suggests to think about someone who has had a limited exposure to birds (Keil calls him Icarus)[15]: because of his limited experience, Icarus bird-specific explanatory knowledge is modest. In addition to knowing many general things about animals that apply to birds as well, Icarus only knows explanatory reasons for properties corresponding to two sorts of bird category contrasts: flightless vs. flying birds and predators vs. prey. Thus, Icarus understands why flightless birds tend to have much smaller wings as compared to body size than flying birds and why their legs tend to be bigger and more muscular; and he understands why predators have many properties that help them catch the prey and why preys have many features that help them avoid predators. In this way explanatory knowledge helps arrange the conceptual space and helps (1) influence intuitive inter-birds similarities, (2) inductions about new properties, (3) and categorization. Moreover, this knowledge provides for help as regards those constraints on correlations between specimen, which are important for the recognition of new elements.

There is, however, a problem with birds that occupy the same spot in this two-dimensional space but which are easily distinguishable in terms of surface feature probabilities and correlations and which Icarus can in fact distinguish, such as a robin vs. a wren. They occupy the same place in Icarus' explanatory matrix and yet Icarus stores additional information about them so as to be able to distinguish them.

15 Keil (1994), 174.

This information may well be largely associative and guided by broad hunches laid down by Icarus' theory of animals and perhaps some general hunches about birds. According to Keil (1994), those hunches insure that even the associatively organized features will not include whether a bird was first seen on a tree with either an odd or an even number of branches, or whether it was on the left or on the right side, or whether it was old or young. But, within those general constraints, many other regularities must indeed be merely stored in terms of correlations and frequencies. In this way our perceptive system is well tuned with the information quality and quantity we are able/have to store in our memory: otherwise we would experience Funes' condition[16].

In some way this could account for the existence of associative structures in concepts. This continuity is both desirable and inevitable simply because we can never have explanations for all regularities that we observe. It is tempting to try to exclude frequency and correlation information from the concept and say that they are only part of heuristics and identification procedures having to do with concept use; but such a move fails to explain why our beliefs are often directed towards explaining those correlations; the correlations form an integral part of the system.

There must be not only explanatory structures that narrow down an indefinitely large number of features and feature relations to a manageable number but also mechanisms for storing information that is outside the ken of current explanation.

According to Keil (1994) there isn't any association-theory *continuum* consisting of strongly increasing and reliable correlations and frequencies. Such a *continuum* could not, however, explain all the cases in which high-numbered correlations and frequencies are insignificant for explaining the belonging to a certain category. Keil gives an interesting example: probably all the washing-machines we saw in our life are white, but the colour/washing-machine correlation doesn't

16 I am referring to the famous tale by Borges, whose main character called Funes, was blindfolded because he received and indelibly remembered every single detail of the objects in his visual field and therefore he was going mad. In Borges (1955).

explain anything about washing-machines. On the contrary the same feature (being white) in polar bears has a causal meaning in species ecology.

6 Conclusions (partial)

In 1998 the debate published on *Cognition* highlighted (1) the importance of the reconsideration of similarity as a cognitive element necessary to many processes concerning concepts and categorization; (2) at the same time the importance of narrowing its field of action by integrating its functions with explanatory and causal aspects that must be borrowed from other theoretic elements: rules, definitions and explanations. Hahn and Chater (1997) analysed the epistemological issue referring to the egg or the hen: what came first the egg or the hen? Are rules the children of similarity or can the latter be explained in terms of rules that underlie any comparison? A final answer is impossible, but at the moment we are able to assume that none of the two aspects can be reduced/brought back to the other. As above said, this brought to the theoretical construction of a model – the Hybrid Model – having many features (flexibility, explanatory power) that fit well human knowledge explanation in its various aspects. And more: is it right to consider the above mentioned Hybrid Model as a 'son' of the Theory-Theory? Is there a sort of continuity between the theoretic claims suggested by Keil, Carey – and generally all the researchers supporting the Theory-Theory – and the arguments supporting this model? Surely these researchers are not able to clearly outline continuity elements in both of them; in other words they assume that the best way for the Theory-Theory real modelization can be a Hybrid Model, but none of them provides for a real relation in their works.

Therefore the partial results are:

1. the acquisition and the conceptual change from an individual theories point of view (in a wide range of knowledge from the first period of our life to adults's 'adult' concepts) cannot be reduced and assimilated tout court to what occurs in scientific theories. From an epistemological point of view the Theory-Theory metaphor is very interesting and evocative but

it shifts the problem without solving it. Moreover, reducing the acquisition and conceptual variation phenomenon we tend to loose important elements featuring those phenomena. In other words, supposing that we can say what concepts really are, we still know so little about the mechanisms underlying acquisition and conceptual change. That remark could be useful from a methodological point of view because phenomena such as, for example, strong and weak restoration give in some cases the possibility to throw light upon these elements[17].

2. Supposing that the Theory-Theory could have a sort of continuity with the Hybrid Model, from a methodological point of view, this model is able to explain the rearrangement processes described according to a shift criterion from the contingent features (/frequency and elements correlation) to the definitions (/rules) highlighting the duality between the theoretical and the atheoretical element of concepts.

Bibliography

G. Boniolo, P. Vidali, *Filosofia della scienza*, Bruno Mondadori Editore, Milano 1999.

J.L. Borges, *Ficciones*, Emecé Editores s.a., Buenos Aires 1955.

S. Carey, *Conceptual Change in Childhood*, MIT Press, Cambridge, MA 1985.

S. Carey, 'Knowledge acquisition: Enrichment or Conceptual Change?', in S. Carey and R. Gelman (eds.), *The Epigenesis of Mind: Essays on Biology and Cognition*, Lawrence Erlbaum Associates, Hillsdale, NJ 1991, pp. 257-291.

L. Celi, *Concepts, Similarity and Theory: towards Hybrid Model*, Degree Thesis in Philosophy, University of Pisa – Department of Philosophy 2001.

M. Chi, R. Glaser, E. Rees, 'Expertise in Problem Solving', in R. Sternberg (ed.), *Advance in the Psychology of Human Intelligence*, vol. 1, Lawrence Erlbaum Associates, Hillsdale, NJ 1982.

B. Cohen, G.L. Murphy, *Models of Concepts*, Cognitive Science, 8, 1984, pp. 27-58.

17 For the Theory-Theory possibilities criticism see Margolis (1995).

P. Feyerabend, 'Explanation, Reduction and Empiricism', in H. Feigl, G. Maxwell (eds.), *Minnesota Studies in Philosophy of Science*, vol. 3, University of Minnesota Press, Minneapolis 1962.

M. Frixione, *Logica, significato e intelligenza artificiale*, Franco Angeli Editore, Milano 1994.

A. Gopnik, 'Conceptual and Semantic Development as Theory Change: The Case of Object Permanence', *Mind and Language*, III, 1988, pp. 197-216.

U. Hahn, N. Chater, 'Concepts and Similarity', in K. Lambert, D. Shanks (eds.), *Knowledge, Concepts and Categories*, MIT Press, Cambridge, MA 1997, pp. 43-92.

J. Hampton, 'Overextension of Conjunctive Concepts: Evidence for a Unitary Model of Concept Typicality and Class Inclusion', *Journal of Experimental Psychology: Learning, Memory, and Cognition*, 14, 1988, pp. 12-32.

J. Katz, *Semantic Theory*, Harper and Row, NY 1972.

F. Keil, 'Conceptual Development and Category Structure', in U. Neisser, *Concepts and Conceptual Development: Ecological and Intellectual Factors in Categorization*, Cambridge University Press, Cambridge 1987, pp. 175-200.

F. Keil, 'Spiders in the Web of Belief: The Tangled Relations Between Concepts and Theories', *Mind and Language*, IV, 1989, pp. 43-50.

F. Keil, 'Explanation, Association, and the Acquisition of Word Meaning', in L. Gleitman, B. Landau (eds.), *The Acquisition of Lexicon*, MIT Press, Cambridge MA 1994, pp. 169-96.

T. Kuhn, *The Structure of Scientific Revolutions*, University of Chicago Press, Chicago 1962.

W. Labov, 'The Boundaries of Words and Their Meanings', in C.J.N. Bailey, R.W. Shuy (eds.), *New Ways of Analyzing Variation in English*, vol. 1, Georgetown University Press, Washington 1973.

G. Lakoff, 'Cognitive Models and Prototype Theory', in U. Neisser (ed.), *Concepts and Conceptual Development: Ecological and Intellectual Factors in Categorization*, Cambridge University Press, Cambridge 1987, pp. 63-100.

E. Margolis, 'The Significance of the Theory Analogy in the Psychological Study of Concepts', *Mind and Language*, 10, 1995, pp. 45-71.

E. Margolis, S. Laurence (eds.), *Concepts: core readings*, MIT Press, Cambridge, MA 1999.

R. Monk, *Ludwig Wittgenstein. The duty of genius*, Johnathan Cape, London 1990.

D. Osherson, E. Smith, 'On the adequacy of prototype theory as a theory of concepts', *Cognition*, 9, 1981, pp. 35-58.

H. Putnam, 'The Analytic and the Synthetic', in H. Feigl and G. Maxwell (eds.), *Minnesota Studies in Philosophy of Science*, vol. 3, University of Minnesota Press, Minneapolis 1962.

W.V.O. Quine, 'Two Dogmas of Empiricism', in *From a Logical Point of View: Nine Logico-Philosophical Essays*, Harvard University Press, Cambridge, MA 1953.

W.V.O. Quine, 'Natural Kinds', in N. Rescher et al., *Essays in Honor of Carl Hempel*, Reidel, Dordrecht 1969, pp. 5-23.

W.V.O. Quine, 'The Nature of Natural Knowledge', in S. Guttenplan (ed.), *Mind and Language*, Oxford University Press, Oxford 1973.

L.J. Rips, 'The Current Status of Research on Concept Combination', *Mind and Language*, 10, 1995, pp. 72-104.

E. Rosch, 'Principles of Categorization', in E. Rosch and B. Lloyd (eds.), *Cognition and Categorization*, Lawrence Erlbaum Associates, Hillsdale, NJ 1978, pp. 27-48.

E. Rosch, C.B. Mervis, 'Family resemblances: Studies in the internal structures of categories', *Cognitive Psychology*, 7, 1975, pp. 573-605.

L. Wittgenstein, *Philosophical Investigations*, Macmillan, NY 1953.

Luciano Celi, Post-graduate Masters in Scientific Communication, International School for Advanced Studies (ISAS), Trieste

Mariella Combi

Cultural models in anthropology

Abstract:

Anthropological models aim to solve the same fundamental problem about the relationships existing between reality, representation and imaginary. It is a characteristic of these models to be always the result of the experience of two continuously present partners: the native and the ethnographer. The creation of different models by natives and ethnographers is the anthropologists' key point, as it is clearly evidenced by the examples which are mentioned here. The aim of explanatory schemata is not limited to render clearer what we observe, but also to construct mental objects so that we may analyse them by comparing the different ensuing representations. So the anthropologist can also construct a sensory model of a society on the basis of the cultural body, as the senses are present in whatever aspect of knowledge.

Keywords:
construction of cultural models, cultural representations, the native and the ethnographer, social structures, sensory order

1 Introduction

Social patterning recalls an activity used unconsciously by us all when we attempt to understand something. In human sciences the creation of models often consists in the reduction of the object's complexity obeying scientific ends. The synthesis and abstraction which also ethnographic models undergo, inevitably entail an impoverishment of 'reality', but they both help to point out the relationships linking the various elements and in finding new connections.

Above all, anthropological models allow one to face the paradoxical situation, described by Franz Boas and reconsidered by Claude

Lévi-Strauss (1952), due to the difference existing between the model constructed by the observer and the model constructed by the observed. It is more difficult to capture the deep structure of a culture in the presence of conscious, self-assured 'local' models, or 'preconceived' interpretative models, of an 'illusory' nature. They induce a deformed interpretation of the data collected by anthropologists.

C. Lévi-Strauss describes the famous example regarding the Bororo of Central Brazil, who are aware of the interlinkage between their marriage rules and the space structure of their villages, divided in two exogamous halves: the members of the eastern clan are supposed to get married to members of the western one. However, the anthropological enquiry discovered a more complex model, not indicated by the Bororo themselves, including the presence of three internal sections in either clan: a 'higher', a 'middle' and a 'lower' one. In addition to the obligation to marry a member of the other clan, both marriage partners should belong to sections at the same level. Sometimes, the informant's unconscious models are more useful than the conscious ones.

Important issues related to the construction of anthropological models focus not only on the formal treatment of the data concerning the subjects/objects of inquiry and the various, possible levels of analysis, but also on the actual meaning one attributes to some concepts, like 'structure' and 'model'. An anthropologist creates a representation of the collected data just as they themselves come out of the object of his research and observation. This characteristic shows how ethnographic models are a product of anthropological theory itself and how data reading is different according to the theoretical model used. This situation created by various interrelated issues is at the origin of the fact that models in anthropology are paraphrased (M. Izard, 1991), except when they apply to those enquiry fields – kinship and sometimes language – for which formalization problems hardly exist.

Another difficulty is due to the fact that the notion of a model designates both a research 'tool' and a research 'object' (U. Fabietti, 1999), since the explanations given by natives about actions and theories concerning their own societies as well as their behaviours are to be regarded as models too. The aim of explanatory schemata is not limited to render clearer what we observe, but also to construct mental objects so that we may analyse them by comparing the different ensuing representations.

2 Social structures, social relations, mental structures

The ambiguity of the ties linking different concepts, sometimes regarded as synonymous, such as 'social structure', 'social relations' and 'cultural model' requires some reflection since many debates have arisen on this subject between the representatives of different theories.

Social structure does not coincide with structural relations and models. According to C. Lévi-Strauss, the notion of a social structure does not refer to empirical reality but to the models constructed upon it. He shares Meyer Fortes' (1949) analysis, in that it is not possible to find a structure in 'concrete reality' because, metaphorically, structure may be defined on the ground of grammar and syntax but not of spoken language.

Nor are 'social relations' synonymous with social structure, as they provide the raw material for the creation of models rendering this structure explicit (C. Lévi-Strauss, 1980). Some anthropologists have made the mistake of analysing the conscious models of indigenous culture as structures whereas they can only be of assistance in discovering the structures of that culture, as in the above example regarding the Bororo.

In addition, structures belonging to different orders are present in a society. For instance, kinship systems have set an order among individuals according to some rules, claims C. Lévi-Strauss, while social organisation is based on different principles and economy on still other precepts. The common characteristic of these examples is to belong all to 'lived orders'; that is, they are functions of an objective reality. They may be faced from outside even without knowing the representations made by the very persons who are being enquired.

It may thus be asserted, together with M. Izard (1991), that with regard to the structural analysis, social relations are the basis for the creation of models putting in evidence social structures. Through a 'transformation' operation it is possible to pass on from one model to another.

On the other hand, A.R. Radcliffe-Brown (1952) writes in *Structure and Function in Primitive Society*:

> When we use the term structure we are referring to some sort of ordered arrangement of parts or components. A musical composition has a structure, and so does a sentence. [...] The components or units of social structure are *persons*, and a person is a human being considered not as an organism but as occupying position in a social structure. [...] The social relationships, of which the continuing network constitute social structure, are not haphazard conjunctions of individuals, but are determined by the social process, and any relationship is one in which the conduct of persons in their interactions with each other is controlled by norms, rules or patterns.

And he correlates structural society with social relationships and behaviour 'so that in any relationship within a social structure a person knows that he is expected to behave according to these norms and is justified in expecting that other persons should do the same'. However, C. Lévi-Strauss, as is well-known, criticised the above author's analogy between organic structure and social structure, between the model of family relations and biological nexus and his failure to distinguish social structure from social relations.

> The structuralism of Lévi-Strauss, in contrast with that of Radcliffe-Brown, looks for social structure not among the observable social relations that obtain in a given culture, but at a deeper level that is revealed only through an analysis of variations at the observable level. [...] The point to stress is that structure is not abstracted from the data but belongs to a model constructed, as it were independently, which under one or another transformation can render immediately intelligible all the observed facts about various domains of social reality. [...] The trouble is that the text last quoted by Lévi-Strauss goes on to describe a structure as a kind of model, rather than a model as a kind of structure. (P. Caws, 1974, p. 2).

Instead, P. Caws maintains:

> If we adopt the definitions of *system* as a set of *entities* mutually interrelated and interdependent, themselves functioning together as an entity at some higher level of organization, and of *structure* as a set of systematic *relations*, concrete if embodied in an actual system, abstract if merely specified but not

> so embodied, then the notion of a *model* can be defined as follows: an abstract structure is a model if it stands for a homologous concrete structure, a concrete structure is a model if it stands for a homologous concrete structure differently embodied. By 'stand for' – continues P. Caws – I mean that features of the model are substituted for features of the structure whose model it is, for purposes of presentation, or instruction, or explanation, or imaginative variation, or computation, or prediction. There need be no preferential direction of the relationship model/modelled: a theory according to one familiar view is a model of the aspects of the world with which it deals, in that we can work out the behaviour of those aspects in theory without having to realize them in practice, in the confidence nevertheless that that is how it would happen in practice; but, on the other hand, a perfectly concrete object (an orrery for example) can be a model for a theory, and so can another theory. To make the structure features of the model central reflects the fact that it stands for the relationship between the entities that constitute the system, rather than for the entities themselves. (P. Caws, 1974, p. 1).

After defining the concepts we are now analysing, P. Caws introduces the notion of a mental structure as an abstract structure connected to language:

> […] as a functioning system depends on *its* embodiment in mental structures. Mental structures, while they have their own autonomy, have evolved as they have because of their modelling relation to structural features of the physical, biological, and social worlds. They are what enables us to respond in the complex ways we do, to remember, to intend, to reflect, to deliberate. But it would be a mistake to say that mental structures model 'real' structures. (P. Caws, 1974, p. 1).

We know that there are structures in the anthropologist's head,

> […] which I have called mental structures […] because they fulfil mental functions (remember that a structure is not a set of entities but a set of relations), and if these structures in turn have the relations of homology and 'standing for' to the social reality with which he deals then they can properly be called models. (P. Caws, 1974, p. 3).

Therefore, there is a relation between the data collected on the field and the model that can be constructed by passing through the connections with social structure and social relations. Several are the models that can be thought of since many are the phenomena to be explained.

3 Some examples of cultural models

On the basis of what has been said so far, the problem of the construction and role of models in anthropology is complex and vaguely delineated and it is chiefly dealing with lineage and alliance theories. The debate between anthropologists on this subject has been very lively as is shown by the number of models proposed on the grounds of different theories. Some of the prevailing models are mentioned here.

3.1 Patterns of culture: Ruth Benedict

Franz Boas, the father and founder of cultural relativism, detects a fundamental problem related to the twofold aspect of models in anthropology. This problem will be reconsidered by Lévi-Strauss. Ruth Benedict, a Boas' pupil, proposes a famous modelling of a psycho-cultural type in *Patterns of Culture* (1934). She says:

> a culture, like an individual, is a more or less consistent pattern of thought and action. Within each culture there come into being characteristic purpose [...] and [...] the heterogeneous items of behaviour take more and more congruous shape [...]. The form that these acts take we can understand only by understanding first the emotional and intellectual mainsprings of that society.

Every culture stands out from the others on account of its tasks and configurations of elements (birth, death, puberty, marriage, beliefs, and so on) interacting between themselves and giving a sense to its particular world. The function of a model here is to integrate coherently, after selection, different cultural traits. The three models Ruth Benedict has selected, which refer to the psychological characteristics of the three cultures she had studied – Apollonian, Dionysian and megalomaniac temperaments – were the basis she departed from to maintain that every person adapts himself or herself to the models which are present in the tradition of his or her community (R. Benedict, 1974).

3.2 Basic or systemic models: Alfred L. Kroeber

Alfred L. Kroeber, another pupil of F. Boas, explains the difference between R. Benedict's models and his own 'basic models'. The former are psychical configurations moulding the typical characteristic of a society so as to impart a certain curvature to the culture of society itself, whereas the 'basic models' are a nexus of cultural traits which have taken on a definite and coherent structure, functioning successfully and acquiring a greater persistence and a heavier historical weight. He analyses the notions of a 'model' and a 'structure', comparing them to their use in biology, on one hand, and introducing the role of history on the other. These basic models have two characteristics: they last a short time and show the possibility of 'cross-breeding' amidst different cultures. An important difference persists, however, between primary, fundamental cultural models – the basic ones which were later on called 'systemic' – and the other secondary models, more superficial and temporary determining which alteration can take place: they are one of the factors which jointly create what occurs (A.L. Kroeber, 1976, p. 93). Clan, totem, potlach, taboo, sacrifice, writing, and so on, are all examples of recurrent and long acknowledged cultural phenomena. In order to find basic models it is necessary, according to Kroeber, to concentrate on the analysis and comparison of correspondences as appropriate for the alphabet in the case of Jewish-Christian-Islamic monotheism, and so on. Basic models represent some inventions and produce important changes.

3.3 Conscious and unconscious, mechanical and statistical models: Claude Lévi-Strauss

Claude Lévi-Strauss points out that both natives and anthropologists devise conscious and unconscious models. This condition does not concern the nature of models. The presence of an unconscious structure makes it possible for the ethnographer to foresee the existence of a model hiding the structure itself. Conversely, conscious models, i.e. 'norms', have the function to perpetuate beliefs.

The construction of a model is based on the observation of facts and on the method selected. In addition, Claude Lévi-Strauss (1980) proposes a distinction between *mechanical* and *statistical models*. He quotes the example of marriage rules in 'primitive' societies (mechanical models) and in ours (statistical models), that is in situations with a different complexity of data. The mechanical model allows one to determine the group of persons who can get married, their social position, their distribution in kinship classes, their future possible partners. A hypothetical model relating to marriage in our societies should be based on statistical data because of the much greater number of unforeseeable factors which should be taken into account, such as the range of possible partners, their social fluidity and so on.

An interesting aspect of structural enquiries is to present structures which can be translated into models with comparable, formal proprieties and which are independent of the elements the structures are made of.

3.4 Segmentary models: Edward Evan Evans-Pritchard

As stated previously, the analyses about the type and role of models in anthropology has chiefly concerned lineage theory, kinship studies which, in the early stages, had as their central concerns the relationship between kinship and territory, and family as a whole. Subsequently, in British anthropology, lineage came to the forefront as a major unit of social organization. In The Nuer (1940) E.E. Evans-Pritchard postulated a model of social structure based on the interaction of territorial and descent principles. The segmentary lineage system, he claimed, functioned in such a way as to define territorial-political units contextually, according to specific situations of opposition and unity and process of fission and fusion. At the same time, the lineage system provided a language in terms of which political relations were expressed and articulated. As a consequence of criticisms, he himself increasingly stressed that the model was not intended to account for actual social organization or group structure on the ground, but rather as an approximation to Nuer values or ideal models of social structure, in terms of which actual social relations could be expressed and interpreted (Seymour-Smith, 1986).

3.5 Explicit and implicit models: Edmund R. Leach

The influence of structuralism led to critiques of 'typologizing' by Edmund Leach, who argued that rather than prejudge the content of social relations by employing the labels 'patrilinity' or 'matrilinity' we should seek more value-free or 'mathematical' ways of expressing the relations between structural elements. He demonstrated the complex interrelationship of ideal models and political action within a historical context in his *Political Systems of Highland Burma* (1954).

He placed great emphasis on the distinction between empirical levels and models as he sustained that the latter are no direct counterpart of reality. According to him 'models are logical construction in the anthropologist's mind' as opposite to natives' models. In *Pul Eliya,* for instance, he defined implicit models as 'jural rules' with which personal behaviour must comply within a society. To these he opposed explicit models, i.e. the statistical norms anthropologists discover concerning the actual persons' behaviour.

3.6 Operational, Representational and Explanatory Models: Peter Caws

Peter Caws suggests a first distinction between 'operational models' and 'representational models': the representational model corresponds to the way the individual thinks things are, the operational model to the way he practically responds or acts (P. Caws, 1974, p. 3). He suggests that it may be helpful to think of events dealt with by the social sciences in terms of the mental structures that determine them (the operational models) and that determine, through reports, one of the sources of our information about them (the representational models) (P. Caws, 1974, p. 4). It can be said that the representational model is conscious and that the operational model is unconscious.

He calls 'explanatory models' those representational models that satisfy criteria of scientific rigor. The explanatory model is always one of the scientists' mental structures; it is never 'in' the empirical data nor (unless the native or the informant also happens to be an anthropologist as well) in the heads of members of the social group being examined. The explanatory model has to take account not only of the empirical data but also of the relationships that are held between these and the operational and representational models in those heads, and failure to this will inevitably result in a misunderstanding of the data (P. Caws, 1974, p. 5).

> The social structure is not identical with the explanatory model, since the latter is in the scientist's head and the former is out there among the people who belong to the society in question. Nor is it identical with the representational or operational models in the natives' heads, since there are many of them and they actually do things – marry and care for one another, use names of special kinds, live in the same place or in different places, and so on. But the structure is dependent in one way or another on these models, since without them the relations that constitute it would not exist. Priority goes, in the end, to the explanatory model, as the only one that is in a position to reflect all the relevant relations and to get them right. (P. Caws, 1974, p. 10).

4 The cultural body: a sensory model

All the above models aim to solve the same fundamental problem about the relationships existing between reality, representation and imaginary. It is a characteristic of these models to be always the result of the experience of two continuously present partners: the native and the ethnographer. The creation of different models by natives and ethnographers is the anthropologists' key point. They have chosen aspects of culture considered more 'concrete' than others: kinship, lineage, alliance, etc. These are 'visible' relationships between individuals, which can be checked, translated into genealogic graphic displays and described through language.

Other representations based on the image of the human body are used by many societies as a model. In a study on the geography of

Madagascar, Chantal Blanc Pomard shows how the natives consider the correspondence between body and landscape both as a model and a mirror. In their view, it is the human body that sets order to the outer world. The perception of bodily coherence helps to structure and codify landscape. References to the body strengthen the role of landscape as a support of information. The space is subdivided by a bodily language which contributes to fix things in the heads and minds of people. The human body provides a model ordering and organizing the space: it enables one to translate the environment. At another level it is possible to think of cultures as contrasting in terms of the distinctive patterns to the 'interplay of the senses' they present. Or the *sensorium*. By the *sensorium* Walter J. Ong means the 'entire sensory apparatus as an operational complex'. He says: 'Given sufficient knowledge of the *sensorium* exploited within specific culture, one could probably define the culture as a whole'. In this case too, formalization is made possible by language because the way the senses are termed and used can reveal a great deal about the sensory model of a culture. We believe that everybody

> divides the sensorium as we do. The Hausa recognize two senses; the Javanese have five senses (seeing, hearing, *talking*, smelling and feeling), [...]. The senses interact with each other first, before they give access to the world; hence, the first step, the indispensable starting point, is to discover what sorts of relations between the senses a culture considers proper. (D. Howes, 1991, p. 258).

The invention of linear perspective vision by Giovanni Battista Alberti in the fifteenth century had a great influence on the *sensorium* of the western culture because, says Robert Romanyshyn, 'what originated with Alberti and his times as a metaphor – look at the world through this grid and it looks like a geometrical pattern – has become for us a map'. The ethnographer risks assuming a similar perspective. The only difference is that it is the 'model of the text' or the 'discursive paradigm' instead of Alberti's window. The visual and 'verbocentric' models of anthropology persist. 'Thus, the shift from the ocular to the oral must be accompanied by a further shift, which takes in the gustatory, olfactory, and tactile modalities as well' (D. Howes 1991, p. 8). The changes in the sensory order must be very clear both in the anthropological theory – with its audio-visual metaphors – and in the construction of a model of the 'cultural body'. The anthropologist must be aware of having a sensory model which is different from the native's one in order to understand a culture.

The anthropologist can construct a sensory model of a culture as the senses are present in whatever aspect of knowledge: because the senses are said, lived, conceived, represented and imagined from cosmology to child-rearing practices, from myth to artefacts, from ritual to proverbs.

A model of the 'cultural body' can be created on the basis of the processes by which its image is shaped; of the representations, symbols and imaginaries of the body itself as they are used in a society; of the connections present in the relations between, body, senses and new technologies. It will also be possible to answer some of the questions asked by the 'Anthropology of senses' of Howes, Classen, Stoller, Seremetakis and so on, such as the following: 'Which senses are emphasized or repressed, and by what means and to which ends?'; 'Is there a natural order of senses?'; 'How is the sensory experience expressed and ordered by language?'; 'Which are the alternatives to our way of perceiving the world?'; 'How does the sensory order relate to the social and symbolic order?'

Bibliography

R. Benedict, *Patterns of Culture*, Houghton Mifflin Company, Boston, New York 1934, It. transl., *Modelli di cultura*, Feltrinelli, Milano 1974[2].

P. Bonte, M. Izard (eds.), *Dictionnaire de l'ethnologie et de l'anthropologie*, PUF, Paris 1991.

P. Caws, 'Operational, Representational, and Explanatory Models', in *American Anthropologist,* 76 (1), 1974.

C. Classen, 'Sweet Colors, Fragrant Songs: Sensory Models of the Andes and the Amazon', in *American Ethnologist*, 17 (4), 1990.

M. Combi, 'The Imagery, The Computer, Artificial Intelligence: a Cultural Anthropological Approach', in *Ai & Society – The Journal of Human Centred Systems and Machine Intelligence*, vol. 6, n. 1, jan-mar, Springer Verlag, London 1992.

M. Combi, *Il grido e la carezza. Percorsi nell'immaginario del corpo e della parola*, Roma, Meltemi 1998^2.

M. Combi, *Corpo e tecnologie. Simbolismi, rappresentazioni e immaginari*, Meltemi, Roma 2000.

M. Combi, 'The cultural body and artificial technology' in M. Negrotti (ed.), *Yearbook of the Artificial*, vol. 1, Peter Lang, Bern, Berlin 2002.

E.E. Evans-Pritchard, *The Nuer: a Description of the Modes of Livelihood and Political Institutions of a Nilotic People*, Oxford University Press, London 1940, It. transl., *I Nuer*, Franco Angeli, Milano 1979.

U. Fabietti, *Antropologia culturale. L'esperienza e l'interpretazione*, Laterza, Roma, Bari 1999.

C. Geertz, *The Interpretation of Cultures*, Basic Book, New York 1973, It. transl., *Interpretazione di culture*, Il Mulino, Bologna 1987.

D. Howes (ed.), *The varieties of Sensory Experience*, University of Toronto Press, Toronto, Buffalo, London 1991.

A.L. Kroeber, *The nature of culture*, The University of Chicago Press, Chicago 1952, It. transl., *Antropologia dei modelli culturali*, Il Mulino, Bologna 1974.

E. Leach, *Political systems of highland Burma. A Study of Katchin Social Structure*, Beacon Press, Boston 1965.

C. Lévi-Strauss, *Anthropologie structurelle*, Plon, Paris 1958, It. transl., *Antropologia strutturale*, Il Saggiatore, Milano 1980.

M. Negrotti, *Naturoids. On the Nature of the Artificial*, World Scientific Publishing, New Jersey, London 2002.

A.R. Radcliffe-Brown, *Structure and Function in Primitive Society*, Macmillan Publishing, New York, London1952.

C. Seymour-Smith, *Dictionary of Anthropology*, The Macmillan Press, London 1986.

Maria Domenica Combi, Dipartimento di Studi romanzi, Facoltà di Scienze Umanistiche, Università 'La Sapienza', Roma

Luciano Floridi and Jeff W. Sanders

The Method of Abstraction

Abstract:

A method is proposed for phenomenological or conceptual analysis, based on making explicit the level of abstraction of discourse. The result formalises an approach that is traditional in science, but has hitherto found little application in philosophy. The constituents of the method are 'observables' collected together and moderated by predicates restraining their 'behaviour'. The resulting collection of sets of observables is called a 'gradient of abstractions' and it formalises the minimum consistency conditions that the chosen abstractions must satisfy.

Two useful kinds of gradient of abstraction – disjoint and nested – are identified. The case is made that in any discrete (as distinct from analogue) domain of discourse a complex phenomenon may be explicated in terms of simple approximations organised together in a gradient of abstractions. Thus the method replaces, for discrete disciplines, the differential and integral calculus, which form the basis for understanding the complex analogue phenomena of science and engineering.

The method is demonstrated on the Turing test, the concept of agenthood, the definition of emergence and the notion of artificial life, and its place in philosophy is considered. It is hoped that our treatment will promote the use of the method in certain areas of the humanities and especially in philosophy.

Keywords:
conceptual scheme, level of abstraction, observables, philosophy of information, Turing test, typed variables

1 Introduction

> If we are asking about wine, and looking for the kind of knowledge which is superior to common knowledge, it will hardly be enough for you to say 'wine is a liquid thing, which is compressed from grapes, white or red, sweet, intoxicating' and so on. You will have to attempt to investigate and somehow explain its internal substance, showing how it can be seen to be manufactured from spirits, tartar, the distillate, and other ingredients mixed together in such and such quantities and proportions. (Gassendi's Fifth Set of Objections to Descartes' Meditations)

Can a complex system always be approximated more accurately at finer and finer levels of abstraction, or are there systems which can simply not be studied in that way? Perhaps the mind, or society, to name two typical examples, are not susceptible to such techniques. That is the defining issue of the recent area of *plectics* [18] and this paper makes no attempt to resolve it. Rather, we clarify that method of approximation at finer levels of abstraction *(the method of abstraction)*, we provide a rigorous formulation of it and we argue for its indispensability in the study of discrete systems.

In studying the characteristics of artificial agents (e.g. [15]) in cyberspace one is presented with a choice: either to acknowledge that such characteristics, including even the properties defining agenthood, are not well defined, or that they depend on the level of abstraction at which they are studied. Although conservativeness would counsel the former, that (negative) approach would be at the expense of a rich theory with potentially interesting applications in the foundations of Information Ethics. So it seems worth exploring the consequences of predicating a theory (of agents, to name but one example) on the level of abstraction. In summary, whether or not a property holds (in particular the definition of agenthood) depends on the level of abstraction of discourse. Sustained study has shown that view not to suffer the flaws of relativism or perspectivism. The purpose of this paper is to present it from scratch, making the claim that, particularly for the study of discrete systems (like cyberspace), the method of abstraction is as important as the method of differential and integral calculus has been for the study of analogue systems.

The paper is methodological and provides applications in various, disparate, areas both conceptual and phenomenological. In the former case it facilitates and clarifies study; in the latter it may also aid synthesis. For in the task of top-down construction, an entity is constructed in several passes, starting from an initial approximation which is subsequently iteratively refined until an acceptable result is produced. That process may be viewed as a succession of exact constructions at progressively less abstract levels of observation. It is in that sense that the method of abstraction underpins top-down construction.

For example an article may be written 'top-down'. The author may determine in advance to start simply with the title. On the second pass the section headings will be decided, presumably reflecting the organisation of material and order of exposition. On the next pass each section is broken into subsections; after that a note is made of the content of each paragraph; and finally sentences are produced. Of course some feedback occurs between passes, but the benefits of that approach are the overall consistency and modularity into components which can be developed independently of each other after a certain stage in the development. Some (hyper) text systems nowadays support that approach to composition. Each pass corresponds to a complete article, at a certain level of abstraction. At the first level of abstraction only the title is visible; and at that level the result of the first pass is indistinguishable from a finished product, since in both cases only the title is observable. At the second level of abstraction the title and the section headings are visible; and so on. There is no difference, in this discrete domain, between the writer's construction of an approximation to the final article at a certain pass and an exact or completed article *at that level of abstraction.* That explains the importance of the method of abstraction for discrete systems. The succession of levels of abstraction we call a 'gradient of abstractions'. That example illustrates the important case we call a 'nested' (or linear) gradient of abstractions.

To ensure the well-definedness of the levels and gradient of abstraction in any application, and to avoid any hint of relativism, some degree of precision is required. Such precision is always possible in discrete systems (where, by definition, only finitely many alternatives are possible for the outcome of each observable), but may not be always possible in analogue systems. That is the reason for the unresolved nature of the fundamental problem of plectics in the case of analogue or hybrid systems, and the reason for our much more positive stance on applicability of the method of abstraction to discrete systems.

If the method of abstraction does indeed have wide application, indications of its use – inchoate at the very least – should be evident in various areas. Part of the reason for the extent of examples in this paper is to follow up that point. Not surprisingly it is common to argue that complex systems are best studied at different levels of analysis (see, for example, [32, 23, 31, 40, 44, 11, 2, 37, 4, 18, 25, 47, 12]). One major success of the general method, from which we have derived inspiration, has been the area of Formal Methods in Computer Science [20, 22]. There, indisputably complex information systems are constructed to meet precise and demanding specifications; and once constructed, existing systems are understood and modified in the on-going task of (software) maintenance. The levels of abstraction are made precise, and at each level a mathematical specification is formulated of the system under construction or study. Techniques are developed and used to ensure that the product at one level of abstraction meets its more abstract specification, and that those techniques are incremental: the validity of the final implementation follows by combining the validity of each incremental synthesis.

We have already mentioned that the fundamental definition of the method of abstraction is that of a 'gradient of abstractions'. Embodied in it are the conditions required for the incremental synthesis, analysis or observation of successively finer levels of abstraction. The definition is founded on the concepts of 'typed variable' (a variable constrained to be of a certain type), 'observable' (an interpreted typed variable), 'level of abstraction' (a collection of observables), 'behaviour' (constrained observable values) and 'moderated level of abstraction' (level of abstraction predicated on behaviour). Those definitions are provided in Section 2, where they are accompanied by a variety of simple examples designed to familiarise the reader with the methodology.

In Section 3 several substantial examples are analysed, to indicate how the method of abstraction may be applied in different areas. We consider the definition of agenthood, the Turing test, the definition of emergence, the definition of artificial life, and we discuss the consistency of the method of abstraction with the notion of quantum observation from Physics, and with the concept of decidable process from Computability.

We are aware that we are proposing a new tool, while philosophers typically attempt to eliminate tools. Quine eliminates the analytic-synthetic distinction; Davidson eliminates conceptual schemes. In Section 4 we make the case for the method of abstraction in the face of concepts like levels of organisation, levels of explanation, conceptual schemes and pluralism, and we confront the problems of relativism and realism.

Although some degree of rigour underlies the definitions on which the method of abstraction is founded, only the rudiments of mathematical notation are presupposed – the main concepts are introduced here without assuming any previous knowledge.

2 Definitions and preliminary examples

In this section we define the simple concepts fundamental to the paper: typed variable, observable, level of abstraction (LoA), behaviour and gradient of abstraction (GoA). Some elementary examples are introduced in order to illustrate the content and variety of the definitions. More elaborate and probing examples are considered in section 3, once the new vocabulary has been established.

2.1 Typed variable

A variable is a symbol that acts as a place-holder for an unknown or changeable referent. A 'typed variable' is to be understood as a variable qualified to hold only a declared kind of data.

> *Definition.* A *typed variable* is a uniquely-named conceptual entity (the *variable*) and a set, called its *type*, consisting of all the values that the entity may take. When notation is required, we shall write x: X to mean that x is a vari-

able of type X. Two typed variables are regarded as *equal* if and only if their variables have the same name and their types are equal as sets. A variable that is not capable of being assigned well-defined values is said to constitute an *ill-typed variable* (see example 2 in section 2.3).

In positing a typed variable an important decision is being made about how its component variable is to be conceived. We shall be in a position to appreciate that fact after the next definition.

2.2 Observable

The notion of an 'observable' is common in science, occurring whenever a (theoretical) model is constructed. The manner in which the features of the model correspond to the situation being modelled is usually left implicit in the process of modelling. In our approach it is important to make that correspondence explicit. We settle on the phenomenological word 'system' to stand for the object of study. This may indeed be what would normally be described as a system in science or engineering, but it may also be a domain of discourse, of analysis or of conceptual speculation.

Definition. By an *observable* we mean an interpreted typed variable, that is, a typed variable together with a statement of what feature of the system under consideration it represents. Two observables are regarded as *equal* if and only if their typed variables are equal, they model the same feature and, in that context, one takes a given value if and only if the other does.

Being an abstraction, an observable is not necessarily meant to result from quantitative measurement or even perception. Although the 'feature of the system under consideration' might be empirical and physically perceivable, it might alternatively be an artefact of a conceptual model, constructed entirely for the purpose of analysis. We are not here concerned with inferential problems associated with the simultaneous observation of several system parameters or of the interaction between the observer and the observed, a point to which we return in section 3.5. (Our approach, in summary, is to model such situations using observables of the kind defined here but constrained appropriately.)

An observable, being a typed variable, has specifically determined possible values. In particular:

> *Definition.* An observable is called *discrete* if and only if its type is finite: it has only finitely many possible values; otherwise it is called *analogue*.

In this paper we are interested in observables as a means of describing behaviour at a quantified (though seldom numerical) level of abstraction; in general several observables will be employed. But before defining the relevant terms, let us consider some simple examples of the two concepts introduced so far.

2.3 Examples

1. Suppose we wish to study some physical human attributes. To do so we, in Oxford, introduce a variable, h, whose type consists of rational numbers. The typed variable h becomes an (analogue) observable once we decide that the variable h represents the height of a person, using the Imperial system (feet and parts thereof).
   To explain the definition of equality of observables, suppose that our colleague, in Rome, is also interested in observing human physical attributes and defines the same typed variable but decrees that it represents height in metres and parts thereof. Our typed variables are the same, but as observables they differ: for a given person the two variables take different representing values. This example shows the importance of making precise the interpretation by which a typed variable becomes an observable.

2. Consider next an example of an ill-typed variable. Were we to be interested in the rôles played by people in some community, we could not introduce an observable standing for those beauticians who depilate just those people who don't depilate themselves (modernising a standard example [43]). For such a variable would

not be well typed, by the standard and straightforward case analysis.
Similarly, each of the standard antinomies [24] reflects an ill-typed variable. Of course the modeller is at liberty to choose whatever type befits the application and if that involves a potential antinomy then the appropriate type might turn out to be a non-well-founded set [3]. But in this paper we shall operate entirely in standard naive set theory.

3. Suppose we were interested in wine. Observables relating to tasting wine include the attributes that commonly appear on 'tasting sheets': 'nose' (representing bouquet), 'legs' or 'tears' (viscosity), 'robe' (peripheral colour), 'colour', 'clarity', 'sweetness', 'acidity', 'fruit', 'tannicity', 'length' and so on, each with a determined type (in spite of a recent trend towards numeric values, these have not been standardised and so we leave to the reader the pleasant task of contemplating appropriate types; for a secondary source of inspiration we refer to tasting-related entries in [41]). If two wine tasters choose different types for, say, 'colour' (as is usually the case) then the observables are different, despite the fact that their variables have the same name and represent the same feature in reality. Indeed, as they have different types they are not even equal as typed variables.
Information about how wine quality is perceived to vary with time – how the wine 'ages' [42] – is important for the running of a cellar. An appropriate observable is the typed variable *a* which is a function associating to each year *y*:*Years* a perceived quality *a*(*y*):*Quality*, where the types *Years* and *Quality* may be assumed to have been previously defined. Thus *a* is a function from *Years* to *Quality*, written $a : Time \rightarrow Quality$. This example demonstrates that types are in general constructed from more basic types, and that observables may correspond to operations, taking input and yielding output. Indeed an observable may be of arbitrarily complex type.

4. The definition of an observable reflects a particular view or attitude towards the entity being studied. Most commonly it corresponds to

a simplification, in which case nondeterminism, not exhibited by the entity itself, may arise. The method is successful when the entity can be understood by combining the simplifications. Let us consider an example.

In observing a game of chess we would expect to record the moves of the game (done by recording the history of the game: move by move the state of each piece on the board is recorded – in English algebraic notation – by rank and file, as are recorded the piece being moved and the consequences of the move). Other observables might include the time taken per move; the body language of the players; and so on. Let us consider, for the sake of argument, a grossly simplified observation.

Suppose we were able to view the chessboard by looking just along *files* (the columns stretching from player to player); we are unable see the ranks or the individual squares. Files cannot sensibly be attributed a colour black or white, but each may be observed to be occupied by a set of pieces (namely those that appear along that file), identified in the usual way (king, queen and so on down to pawn). With this view a move may be observed by the effect it has on the file of the piece being moved. For example a knight moves one or two files either left or right from its starting file. A bishop is indistinguishable from a rook which moves along a rank; and a rook which moves along a file appears to remain stationary. Whether or not a move results in a piece being captured, appears to be nondeterministic.

Whilst the 'underlying' game is virtually impossible to reconstruct, each state of the game and each move (i.e. operation on the state of the game) can be 'tracked' with this dimensionally impoverished family of observables. If a second view is then taken corresponding instead to rank, and the two views combined, then the original game of chess can be recovered (since each state is determined by its rank and file projections, and similarly for each move). The two disjoint observations together reveal the underlying game.

5. The degree to which a type is appropriate depends on its context and use. For example, to describe the state of a traffic light in Rome we might decide to consider an observable *colour* of type

{red, amber, green} that corresponds to the colour indicated by the light. That option abstracts the length of time for which the particular colour has been displayed, the brightness of the light, the height of the traffic light, and so on. That is why the choice of type corresponds to a decision about how the phenomenon is to be regarded. To specify such a traffic light for the purpose of construction, a more appropriate type would comprise a numerical measure of wavelength. In Oxford, the type of colour would be a little more complex, since – in addition to red, amber and green – red and amber are displayed simultaneously for part of the light's cycle. So, an appropriate type would be *{red, amber, green, red-amber}*.

6. The design of a database is a special case of the definition of a collection of observables. In a database an observable is called a key, its correspondence with reality is left implicit (though it is usually reflected in the name) and its type is inferred from either the values it takes or from its declaration in the database programming language. The type 'finite string of characters' is frequently used, often being the most appropriate concrete method of description. Examples include names, addresses and such like.
   So too with an observable in general: it is sometimes preferable not to provide in advance all the possible outcomes (i.e. a type) but simply to define its type to consist of all finite sequences of characters, with each value equal to a character string. That definition reflects a decision that, although the observable is well-typed, its actual type is not of primary concern.

7. The method proposed in this paper will seem familiar not only to scientists. With the passing of the classical view that a literary text has *the* absolute view determined by its author, deconstruction promotes the revealing of other, sometimes implicit, values embodied in a text. As Fish ([13], page 327) writes:

   > Interpretation is not the art of construing but the art of constructing. Interpreters do not decode poems; they make them.

In that art the interpreter's view is determined by a collection of observables, documented in order to make explicit a view taken. To consider a popular example [49], when interpreting the line

> A rose is a rose is a rose

we might choose observables that relate to the semantics of the text, or that reflect syntax only, the rhythmic nature of the words. We might further choose observables that permit the reader to make a comparison with the rhythms of a Flamenco dancer, and so forth.

Thus a reading of Shakespeare might be based on observables that relate to literary style; or to history; or economics; or social (in particular gender) issues; or theatrical concerns; and so on. A collection of observables provides a particularly clear way of acknowledging and documenting the different views and how they interact. Part of the early appeal of deconstruction [45, 46] was that, by revealing values implicit in the text, interesting behaviour appeared as emergent (in a sense to be made explicit later): it required the discovery of new observables to make it explicable.

It is hardly surprising that the use of observables has powerful application in textual analysis, in view of its close relationship with the science of anthropology:

> Doing ethnography is like trying to read (in the sense of 'construct a reading of') a manuscript [...] [17]

2.4 Level of Abstraction, LoA

Any collection of typed variables can, in principle, be combined into a single 'vector' observable whose type is the Cartesian product of the types of the constituent variables. In the wine example, the type *Quality* might be chosen to consist of the Cartesian product of the types *Nose*, *Robe*, *Colour*, *Acidity*, *Fruit* and *Length*. The result would be a single, more complex, observable.

In practice, however, such vectorisation is unwieldy since to express a constraint on just some of the observables would require projection notation to single out those observables from the vector. Instead we base our approach on a *collection* of observables. Since the very choice of a collection of observables characterises an approach to some phenomenon, it warrants a name.

> *Definition.* A *level of abstraction, LoA,* is a finite but non-empty set of observables. No order is assigned to the observables, which are expected to be the building blocks in a theory characterised by their very definition. A LoA is called *discrete* [respectively *analogue*] if and only if all its observables are discrete [respectively analogue]; otherwise it is called *hybrid.*

In considering wine, different LoAs are appropriate for different purposes. To evaluate a wine the 'tasting LoA', consisting of observables like those mentioned in the previous section, would be relevant. For the purpose of ordering wine, a 'purchasing LoA' would be appropriate, containing observables like 'maker', 'region', 'vintage', 'supplier', 'quantity', 'price', and so on; but here the 'tasting LoA' would be irrelevant. For the purpose of storing and serving wine, the 'cellaring LoA' would be relevant, containing observables for 'maker', 'type of wine', 'drinking window', 'serving temperature', 'decanting time', 'alcohol level', 'food matchings', 'quantity remaining in the cellar', and so on.

The traditional sciences tend to be dominated by analogue LoAs, the humanities and information science by discrete LoAs and mathematics by hybrid LoAs. We are about to see why the resulting theories are fundamentally different.

2.5 Behaviour

The definition of observables is only the first step in studying a system at a given level of abstraction. The second step consists of deciding what relationships hold between the observables, a task we support in this section by introducing the concept of system 'behaviour'. We shall see that it is the fundamentally different ways of describing behaviour in analogue and discrete systems which accounts for the differences in the resulting theories.

Not all values exhibited by combinations of observables in a LoA may be realised by the system being modelled. For example if the four traffic lights at an intersection are modelled by four observables, each representing the colour of a light, the lights can not in fact all be green together. In other words, the combination in which each observable is green cannot be realised in the system being modelled, although the types chosen allow it. Similarly the choice of types corresponding to a rank-and-file description of a game of chess allow the white queen and the white king to be placed on any square, whereas in fact they cannot occupy the same square.

Some technique is therefore required to describe those combinations of observable values that are allowed. The most general method is simply to describe the allowed combinations of values. Such a description is determined by a predicate, whose allowed combinations of values we call the 'system behaviours'.

> *Definition.* A *behaviour* of a system at a given LoA is defined to consist of a predicate whose free variables are observables at that LoA. The substitutions of values for observables that make the predicate true are called the *system behaviours*. A *moderated* LoA is defined to consist of a LoA together with a behaviour at that LoA.

To pursue the previous examples, in reality human height does not take arbitrary rational values: it is positive and bounded above by (say) nine feet. The variable h representing height is therefore constrained to reflect reality by defining its behaviour to consist of the predicate $0<h<9$, in which case any value of h in that interval is a 'system' behaviour.

Wine is not realistically described by arbitrary combinations of the observables mentioned above. For example it cannot be both white and highly tannic. As an example of the type of character strings, valid addresses on letters posted in England might be decreed to end either in the name of a country or in a UK postcode. That predicate determines which character strings are permitted behaviours of character strings representing addresses.

Since Newton and Leibniz, the behaviours of the analogue observables studied in science have typically been described by differential equation. A small change in one observable results in a small, quantified, change in the overall system behaviour. Accordingly it is the rates at which those smooth observables vary which is most conveniently described[1]. The desired behaviour of the system then consists of the solution of the differential equation. But that is a special case of a predicate: the predicate holds at just those values satisfying the differential equation. If a complex system is approximated by simpler systems, then the differential calculus provides a supporting method for quantifying the approximation.

The use of predicates to demarcate system behaviour is *essential* in any (nontrivial) analysis of discrete systems, for the following reason. In discrete systems no such continuity holds: the change of an observable by a single value may result in a radical and arbitrary change in system behaviour. Yet complexity demands some kind of comprehension of the system in terms of simple approximations. When that is possible, the approximating behaviours are described exactly, by a predicate, at a given LoA and it is the LoAs that vary, becoming more comprehensive and embracing more detailed behaviours until the final LoA accounts for the desired behaviours. Thus the formalism provided by the method of abstraction can be seen as doing for discrete systems what differential calculus has traditionally done for analogue systems.

The use of predicates is essential in subjects like Information Science where discrete observables are paramount and hence predicates are required to describe behaviour. In particular, state-based methods like Z [48, 20] provide notation for structuring complex observables and behaviours in terms of simpler ones. Their primary concern is with the syntax for expressing those predicates, an issue we avoid where possible in this paper by stating predicates informally.

It is time now to combine approximating moderated LoAs to form the primary concept of the method of abstraction.

1 It is interesting to note that the catastrophies of *chaos theory* are not smooth, although they do appear so when extra observables are added, taking the behaviour into a smooth curve on a higher-dimensional manifold. Typically, chaotic models are weaker than traditional models, their observables merely reflecting *average* or *long-term* behaviour. The nature of the models is clarified by making explicit the LoA.

2.6 Gradient of abstractions, GoA

For a given (phenomenological or conceptual) system or feature, different LoAs correspond to different representations or views. A 'GoA' is a formalism defined to facilitate discussion of discrete systems over a range of LoAs. Whilst a LoA formalises the scope or granularity of a single model, a GoA provides a way of varying the LoA in order to make observations at differing levels of abstraction. For example, in evaluating wine we might be interested in the GoA consisting of the 'tasting' and 'purchasing' LoAs, whilst in managing a cellar we might be interested in the GoA consisting of the 'cellaring' LoA together with a sequence of annual results of observation using the 'tasting LoA'.

In general, the observations at each LoA must be explicitly related to those at the others; to do so we use a family of relations between the LoAs. For this we need to recall some (standard) preliminary notation.

> *Notation.* A *relation* from a set A to a set C is a subset of the Cartesian product $A \times C$. It is thought of as relating just those pairs (a,c) that belong to the relation. The *reverse* of relation R is its mirror image: $\{(c,a) \mid (a,c) \in R\}$. A relation R from A to C translates any predicate p on A to the predicate $P_R(p)$ on C which holds at just those $c{:}C$ which are the image through R of some $a{:}A$ satisfying p
>
> $$P_R(p)(c) \;=\; \exists a{:}A \;\; R(a,c) \wedge p(a)$$

The translation P_R, with its dual partner translating in the opposite direction (not required here), was introduced by Peirce [39].

We have finally come to the main definition of the paper.

> *Definition.* A *gradient of abstractions, GoA*, is defined to consist of a finite set[2] $\{L_i | 0 \le i < n\}$ of moderated LoAs L_i and a family of relations $R_{i,j} \subseteq L_i \times L_j$, for $0 \le i \ne j < n$, relating the observables of each pair L_i and L_j of distinct LoAs in such a way that:
>
> 1. the relationships are inverse: for $i \ne j$, $R_{i,j}$ is the reverse of $R_{j,i}$

2 The case of infinite sets has application to continuous systems but is not considered here.

2. the behaviour p_j at L_j is at least as strong as the translated behaviour $P_{R_{i,j}}(p_i)$

$$p_j \Rightarrow P_{R_{i,j}}(p_i) \qquad (1)$$

Two GoAs are regarded as *equal* if and only if they have the same moderated LoAs (i.e. the same LoAs and moderating behaviours) and their families of relations are equal. A GoA is called *discrete* if and only if all its constituent LoAs are discrete.

Condition (1) simply means that the behaviour moderating each lower LoA is *consistent* with that specified by a higher LoA. Without it the behaviours of the various LoAs constituting a GoA would have no connection to each other. A special case, to be elaborated in the definition of 'nestedness' to follow, might clarify the point. If one LoA L_i extends another L_j by adding new observables, then the relation $R_{i,j}$ is the inclusion of the observables of L_i in those of L_j and (1) reduces to: the constraints imposed on the observables at LoA L_i remain true at LoA L_j, where 'new' observables lie outside the range of $R_{i,j}$.

A GoA whose sequence contains just one element evidently reduces to a single LoA. So our definition of 'LoA' is subsumed in that of 'GoA'.

The consistency conditions imposed by the relations $R_{i,j}$ are in general quite weak. It is possible, though of little help in practice, to define GoAs in which the relations connect the LoAs cyclically. Of much more use are these two important kinds of GoA: 'disjoint' GoAs (whose views are complementary) and 'nested' GoAs (whose views provide successively more information). Before defining them we need a little further notation.

Notation. We recall that a *function f* from a set C to a set A is a relation, i.e. a subset of the Cartesian product $C \times A$, which is single-valued

$$\forall c{:}C\ \forall a,a'{:}A\ ((c,a) \in f \wedge (c,a') \in f) \Rightarrow a = a'$$

(which property means that the notation $f(c)=a$ is a well-defined alternative to $(c,a) \in f$) and total

$$\forall c{:}C\ \exists a{:}A\ f(c)=a$$

(which means that $f(c)$ is defined for each $c{:}C$). A function is called *surjective* if and only if every element in the target set lies in the range of the function:

$$\forall a{:}A\ \exists c{:}C\ f(c)=a$$

Definition. A GoA is called *disjoint* if and only if the L_i are pairwise disjoint (i.e. taken two at a time, they have no observable in common) and the relations are all empty. It is called *nested* if and only if the only nonempty relations are those between L_i and L_{i+1}, for each $0 \leq i < n-1$, and moreover the reverse of each $R_{i,i+1}$ is a surjective function from the observables of L_{i+1} to those of L_i.

A disjoint GoA is chosen to describe a system as the combination of several non-overlapping components. That is useful when different aspects of the system behaviour are helpfully thought of as being determined by the values of distinct observables. This case is rather simplistic since more typically the LoAs are tied together by common observations. For example the services in a domestic dwelling may be represented by LoAs for: electricity, plumbing, telephone, security and gas. Without going into detail about the constituent observables, it can be seen that in an accurate representation, the electrical and plumbing LoAs would overlap (why?) whilst the telephone and plumbing would not.

A nested GoA is chosen to describe a complex system exactly at each level of abstraction and incrementally more accurately. The condition that the functions be surjective means that any abstract observation has at least one concrete counterpart. As a result the translation functions cannot overlook any behaviour at an abstract LoA: behaviours lying outside the range of a function translate to the predicate *false*. The condition that the reversed relations be functions means that each observation at a concrete LoA comes from at most one observation at a more abstract LoA (although the converse fails in general, allowing one abstract observable to be refined by many concrete observables). As a result the translation functions simplify.

For example, the case of a traffic light which is observed to have colour *colour* of type *{red, amber, green}* is captured by a LoA, L_0, having that single observable. If we wish to be more precise about colour, perhaps for the purpose of constructing a traffic light, we might consider a second LoA, L_1, having the variable *wl* whose type is a positive real number corresponding to the wavelength of the colour. To determine the behaviour of L_1, suppose that constants $\lambda_{red}<\lambda_{red'}$ delimit the wavelength of red, and similarly for amber and green. Then the behaviour of L_1 is simply this predicate with free variable *wl*

$$(\lambda_{red}\leq wl\leq\lambda_{red'})\vee(\lambda_{amber}\leq wl\leq\lambda_{amber'})\vee(\lambda_{green}\leq wl\leq\lambda_{green'}).$$

Then the sequence consisting of the LoA L_0 and the moderated LoA L_1 forms a nested GoA. Informally, the smaller, abstract, type *{red, amber, green}* is a projection of the larger. The relevant relation associates to each value *c:{red, amber, green}* a band of wavelengths perceived as that colour. Formally, *R(colour,wl)* is defined to hold if and only if for each *c:{red, amber, green}*,

$$colour=c \text{ iff } \lambda_c\leq wl\leq\lambda_{c'}.$$

In the wine example, a first LoA might be defined to consist of the variable 'kind' having type red, white, rose under the obvious representation. A second LoA might be defined to consist of the observable 'kind' having type

{stillred, sparklingred, stillwhite, sparklingwhite, stillrose, sparklingrose}.

Though the second type does not contain the first, it produces greater resolution under the obvious projection relation. Thus the GoA consisting of those two LoAs is nested.

Those two important forms of GoA – disjoint and nested – are in fact theoretically interchangeable. For if *A* and *B* are disjoint sets then *A* and their union $A\cup B$ are increasing sets and the former is embedded in the latter; thus a disjoint GoA can be converted to a nested one. Conversely if *A* and *B* are increasing sets with the former embedded in the latter, then *A* and the set difference $A \setminus B$ are disjoint sets; thus a nested GoA can be converted to a disjoint one.

Following the technique used to define a nested GoA, it is possible to define less restricted but still hierarchical GoAs. Important examples include tree-like structures, of which our nested GoAs are a special, linear, case.

For theoretical purposes, the information captured in a GoA can be expressed equivalently as a single LoA of more complicated type (namely one whose single LoA has type equal to the sequence of the LoAs of the complex interface). However the current definition is better suited to application.

2.7 Methodology

Models are the outcome of the analysis of a system, developed at some LoA(s) and from some perspective. An important contribution of these ideas is to make precise this commitment to a LoA/GoA before further elaborating a theory. We call this the *method of abstraction*. The benefits of this method are twofold.

On the one hand, specifying the LoA means clarifying from the outset the range of questions that (a) can be meaningfully asked and (b) are answerable in principle. One *might* think of the input to a LoA as consisting of the system under analysis, comprising data; its output is a model of the system, comprising information [52]. The quantity of information in a model varies with the LoA: a lower LoA, of greater resolution or finer granularity, produces a model that contains more information than a model produced at a higher, or more abstract, LoA. Thus, a given LoA provides a quantified commitment to the kind and amount of information that can be 'extracted' from a system. The choice of a LoA pre-determines the type and quantity of data that can be considered and hence the information that can be contained in the model. So knowing at which LoA the system is being analysed is indispensable, for it means knowing the scope and limits of the model being developed.

On the other hand, being explicit about the LoA provides a good antidote against ambiguity and misunderstanding. The risk without it is that the writer may be writing with one LoA in mind whilst the reader may be using another. Or that one analysis may use a certain kind of multi-levelled approach incompatible with another. The examples of the previous section make this point.

3 Case studies

We shall now demonstrate the advantages of the method of abstraction by applying it to some more substantial examples.

3.1 Agents

An *agent* can be thought of [15] as a *transition system* (i.e. a system of states and transitions between them) that is *interactive* (i.e. responds to stimulus by change of state), *autonomous* (i.e. is able to change state without stimulus) and *adaptable* (i.e. is able to change the transition rules by which it changes state). However each of those properties, and hence the definition of agenthood, makes sense only at a prescribed LoA.

For example whether or not a rock is deemed to be interactive depends on the length of time and level of detail of observation. Over a long period it erodes and hence changes state. At day it absorbs solar radiation which it emits at night. But with observables resulting from scrutiny over a period of ten seconds by the naked eye from ten metres, it can be deemed not to be interactive.

If the LoA at which one observes it abstracts gravity and resistance, a swinging pendulum appears to be autonomous but neither interactive nor adaptive. By extending the LoA to incorporate air resistance, it becomes adaptive. By observing also the whistling sound it makes with the air, it becomes interactive.

If a piece of software that exhibits machine learning [33] is studied at a LoA which registers its interactions with its environment, then the software will appear interactive, autonomous and adaptive, i.e. to be an agent. But if the program code is revealed then the software is shown to be simply following rules and hence not to be adaptive. Those two LoAs are at variance. One reflects the 'open source' view of software: the user has access to the code. The other reflects the commercial view that, although the user has bought the software and can use it at will, he has no access to the code. At stake is whether or not the software forms an (artificial) agent. For further examples we refer to [15].

3.2 The Turing test

Turing [53] took the incisive step of arguing that the ability to think (since called 'intelligence') can be satisfactorily characterised by means of a test, rather than by explicit definition. In retrospect, that step may seem a small one. After all, we are quite familiar with areas in which no explicit definition is possible or sensible. Society makes no attempt to characterise what it means to be an acceptable driver in terms of vision, response times, coordination, experience and other physical attributes. Instead it relies on a driving test. Society does not attempt to define what it means for a school student to have reached an acceptable academic standard by the end of school; it relies on final school examinations. But incisive that step certainly must have been in view of the vast number of attempts (even to this day) to characterise intelligence explicitly.

Opponents of Turing's approach usually object that his test functions at the wrong level of abstraction: perhaps it ought to include a component of creativity, of spontaneity, and so on. But without concepts like those introduced above, it is very difficult to make one's objections precise or defend Turing's approach. It is therefore of considerable interest to see, first, how the Turing test can be expressed using phenomenological LoAs and, second, how it can be analysed using conceptual LoAs.

3.2.1 Turing's imitation game

Let us start, as did Turing, by considering an imitation game, in which a man A and a woman B are placed in a room separate from an interrogator C, who communicates with each by teleprinter (these days replaced by computer). C puts questions to A and B, known only as X and Y. C's task is to identify $X=A$ and $Y=B$ or, conversely, $X=B$ and $Y=A$, by considering their responses.

We might describe that scenario by taking a first, extremely abstract, LoA to reflect just the correctness of *C*'s identification. The LoA L_0 consists of a single variable *ans* of type *{right, wrong}* which becomes an observable under the correspondence: *ans* takes the value *right* if *C* is correct and the value *wrong* if *C* is incorrect. In choosing that LoA, we are intentionally abstracting the actual answer (whether *X* was *A* or *B*), the questions and answers, response times, and so on in order to capture just the outcome of the imitation game.

We might reveal *C*'s actual identification by defining a second, disjoint, LoA L_1 whose single variable, *Z*, is of type *{(A,B), (B,A)}* which is made into an observable under the correspondence that the first component of *Z* is the putative identity of *X* and the second component that of *Y*. Combining the two LoAs L_0 and L_1 gives a disjoint GoA.

Of course there are alternative approaches, which is why it is important to be precise about the one taken. We might have defined a GoA by replacing the LoA L_1 with a LoA containing two observables, the first corresponding to the identity of *X* and the second corresponding to the identity of *Y*. That would be more involved, since each would have type *{A, B}* and we would have to moderate it with the behaviour that the values of *X* and *Y* differ. Our choice of L_1 avoids the complication by building that behaviour into the type of *Z*. But with several observables, in general such moderating behaviours cannot be avoided.

To model *C*'s questions, the addressees and their responses, we define a third LoA, L_2. Let *Q* and *R* denote the sets of possible (well-posed) questions and responses respectively (an example where the type of text strings may be considered appropriate). Then each 'question, addressee and response' triple is a variable whose type is the Cartesian product of *Q*, *{X, Y}* and *R*. It becomes an observable under the correspondence just established. The observable we seek now consists of a sequence (of arbitrary but finite length) of such triples, corresponding to the sequence of interactions in temporal order; and L_2 contains that single observable. (An alternative would be to have an observable for the number of questions, an observable for each question and an observable for each response.) L_2 can be added to either GoA to obtain a GoA which is still disjoint but has higher resolution.

More detailed LoAs are possible and easy to define but, following Turing, we stop here having appreciated that any discussion of the imitation game may be accurately 'calibrated', according to its level of abstraction, with a GoA.

3.2.2 Turing's test

In the Turing test, A is replaced by a 'machine' (nowadays 'computer'). Turing proposed that the question 'Can machines think?' be replaced by the question: 'Will the interrogator decide wrongly as often when the game is played like this as he does when the game is played between a man and a woman?'.

These days[3] the test is normally stripped of its sex-specific nature and the interrogator is simply asked to determine the human from the machine. Appropriate GoAs are defined as above, but with A representing a computer and B a human.

Although Turing did not make it explicit, the phrase 'as often' in his description implies repetition of the test, and a conclusion reached by statistical analysis. Suppose that C initiates a pair of question/answer sessions of the type used in the imitation game. A list of questions is put to two situations, one containing a man and a woman, the other containing a machine and a woman. We suppose that the answers in the first situation are of type A_1 and those in the second of type A_2, thus avoiding here the question of whether or not $A_1=A_2$. As before, C makes an identification. The appropriate LoA has type, call it J, equal to the Cartesian product of the type $\{(A,B), (B,A)\}$ and the type of all sequences of elements of the type $Q \times \{X, Y\} \times A_1 \times A_2$.

The observable corresponding to repetition of that situation j times, though not necessarily with the same questions, has type consisting of the Cartesian product of j-many copies of type J, namely J^j. The LoA incorporating that observation plus the answer to the ultimate question is then the Cartesian product of the type J^j and the type $\{right, wrong\}$.

3 For a summary of the Turing test today, and its incarnation in competitive form (the Loebner prize), see [34].

Likewise, a more complex type can be constructed to reveal the nature of the statistical test; in this case too, we follow Turing and overlook the details.

3.2.3 Turing discussed

The previous two sections have shown how to formalise the Turing test using phenomenologically motivated GoAs. But the method of abstraction can be used to discuss and compare variations on the test. Indeed it is difficult to imagine how such an analysis could be formulated without a concept equivalent to LoA. Now details of the LoAs need not be given as long as it is clear that they could be.

Turing couched his test in terms of a single human interrogator. In the Loebner test [34] interrogation is provided by a panel of humans interacting via computer interface in real time. Alternatively the interrogator could be a machine; or instead of real-time interaction, a list of pre-arranged questions might be left, and the list of answers returned to the interrogator; or instead of serial questions and answers the interrogator might hold a 'general conversation' [34]. Each alternative modifies the power of the test. All can be formalised by defining different GoAs.

The Turing test might be adapted to target abilities other than 'thinking', like: interpretation of text, game playing, puzzle solving, spatial reasoning, creativity, and so on. Contemplating the possible GoAs provides a way to formalise the variant test clearly and elegantly, and promotes simple comparison with contending treatments.

3.3 Emergence

The method of abstraction is ideally suited to the study of systems so complex that they are best understood stepwise: by their gradual disclosure at increasingly fine levels of abstraction. The study of such systems strikes at the heart of the controversy mentioned in the introduction: how does ultimate complexity arise in the stepwise approximation by simple systems? Gell-Mann [18] has suggested that the study of such phenomena be called *plectics*, and introduces it using an idea he calls *granularity* which is conveniently formalised by LoA.

A key concept in such an approach to complex systems is that of 'emergent behaviour': behaviour that arises in the move from one LoA to a finer level (see [21], section 2 for a discussion). In this section, we apply the method of abstraction to clarify that concept of emergence; in the next we go on to summarise the method's application to one particular area in which emergence has received much attention, that of artificial life (ALife).

3.3.1 Tossing a coin

Emergence, not surprisingly for such an important concept, can take various forms [36, 9]. It derives from the notion that, according to Hendriks-Jansen ([21], page 283),

> properties at higher levels are not necessarily predictable from properties at lower levels.

There 'lower levels' are more abstract and 'higher levels' more detailed or concrete. We are concerned here only with that idea of emergence, which is neatly captured using a GoA containing two nested LoAs. We begin with an example.

The process of tossing a coin may be modelled abstractly with an observable *outcome* of type *{head, tail}* corresponding to the side of the coin that faces upwards after the toss. This LoA abstracts any other result (like the coin's landing on its edge or becoming lost) and other features like manner of tossing, number of spins, time taken and so on. In particular it models just one toss of the coin and so cannot account for the 'fairness' of the coin, which would reveal itself statistically after a large number of tosses.

Now suppose we wish to model the repetition of that process with the explicit aim of discussing a coin's fairness. We introduce a more concrete LoA whose observables are: a natural number n corresponding to the number of tosses of the coin; and a list of n values from *{head, tail}* corresponding to successive tosses as modelled above. At this second LoA we are able to make judgements (using standard statistics, for example) about the fairness of the coin, based on the frequency of the outcomes.

This example demonstrates emergence as follows. In many repeated tosses of the coin the more abstract model applies toss by toss, but does not allow frequency of outcome to be observed, as it is in the finer model. We say that the notion of the coin's fairness is emergent at the finer LoA.

3.3.2 Emergence defined

That situation can be formalised as follows. Suppose some system is under consideration using a nested GoA consisting of two LoAs. Suppose the more abstract LoA, *A*, is moderated by behaviour p_A describing the abstract view of system and the more concrete LoA, *C*, is moderated by behaviour p_C describing the concrete view. The abstract and concrete observables are related by the total and one-to-many relation $R_{A,C}$. Recall that a behaviour of the system at LoA *C* is a tuple of values for the observables of *C* that satisfies p_C.

> *Definition.* A behaviour of the system at LoA *C* is said to be *emergent* (with respect to that nested GoA) if and only if its translation under the relation $R_{A,C}$ fails to satisfy p_A. *Emergence* is said to hold in the GoA if and only if there is some emergent behaviour.

There is frequently confusion about emergence. This is scarcely surprising, since without the notion of LoA the various levels at which a system is discussed cannot be formalised.

Emergence arises typically because the concrete LoA embodies a 'mechanism', or rule for determining an observable, which has been overlooked at the abstract LoA, usually quite deliberately in order to gain simplicity at the cost of detail. Frequently the breakthrough in understanding some complex phenomenon has come by accounting for emergent behaviour; and that has resulted from considering the process by which it occurs, rather than taking a more static view of the ingredients involved. In the coin example we have avoided incorporating any mechanism; but any of the multitude of pseudo-random number generators could be used to generate lists of *head* and *tail* and hence to account for the emergent phenomenon [26].

An interesting example of emergence is provided by quantum mechanics, according to which each action (other than the process of observation) is (unitary and hence) reversible. Yet when observations 'in the large' are made of huge physical systems, in spite of the components of those systems obeying the laws of quantum mechanics, the laws of thermodynamics *emerge*: in spite of local reversibility, entropy increases.

3.3.3 Process observed

The majority of observables considered have been 'static'. However operations constitute vital observables. Indeed, the importance of 'process' may be indicated by the example of a sponge cake. With only the ingredients as observables (i.e. the amount of each ingredient) the sponge-like nature of the cake is, as many a novice cook has found, emergent. But if the manner of aeration (a variable indicating the aeration effect of bicarbonate of soda under the right conditions) is also an observable then sponginess is explicable. In other words, the behaviour of sponginess emerges at the finer level of abstraction only.

3.4 Artificial life

Attempts to define artificial life (ALife) have been no more successful than the attempts, both before and after Turing, to define 'intelligence' explicitly. Having chosen a LoA *L*, we propose that ALife with respect to *L* be defined to consist of all those entities observably indistinguishable under *L* from a live entity. A sensible 'sanity check' on *L* is that the result does not depend on the particular species of life observed, provided that it is chosen with regard to the concept of life being explored.

To give a rather coarse example, perhaps the most abstract observation possible of a population is its size: the number of members as a function of time. That is an observable of type 'whole number' and it is well-typed provided that the population is finite and itself well defined. At this LoA a population provides no more information than its size; there is no distinction between live populations, artificial populations and for that matter any set which varies in size. However such a LoA is sufficient to support the well known Fibonacci model of population size per generation ([51], page 923). It is already of some interest if one adds observables including the rates of birth, mortality, migration and harvesting. It then becomes sufficient, for example, to discuss age distribution across the population and for the management of pest populations, for the harvesting of natural populations, for the modelling of insect outbreaks, of interacting populations and so on [35]. Of course at this LoA an observer from space might decide that cars are the dominant life form on parts of earth ([1], page 13).

A slightly finer LoA is that at which the location of individual members of a population is plotted against time (say in Oxford, every 24 hours). At this 'spatial distribution' LoA humans are indistinguishable from ants, bicycles and books: all move around, are created and destroyed. (Were it not for copyright regulations, books might also be 'cloned').

The method of abstraction seems to apply well to summarise, clarify and facilitate the comparison of existing contributions [16] to the theory of Alife. One important approach has been that of Langton [29] which extends the biological notions of genotype and phenotype to gtype and ptype respectively so that those terms apply equally to ALife. Gtype refers to a low-level implementation mechanism, behaviour or LoA whilst a ptype refers to a higher-level behavioural structure or LoA that emerges as a result of those mechanisms interacting. Langton, [29] section 5, discusses the subtleties of inferring the latter from the former. His model provides an important instance of the use of a GoA containing two interfaces, one for gtype observables and the other for ptype observables.

3.5 Quantum observation

It is remarkable that the disparate disciplines of quantum mechanics and social anthropology share a fundamental feature: in each, observation automatically involves interference. Observing a quantum or anthropological system is possible only at the expense of a change to the system. By contrast our definition of observable makes no assumptions about how the entity is (capable of being) measured (effectively) in practice. In this section we address that issue.

In quantum mechanics 'observable' means something much more restricted than the sense in which we have used it here. There it is to be distinguished from state that is posited to exist in order to explain the frequency with which observables take their values. Such 'beables' [5] are for us also observables as is, for that matter, the frequency with which an observable takes on its values. The latter might be regarded as unachievable in practice since any finite number of readings can achieve only an approximation to it. But that need be of no concern to us. Our only requirement is that an observable be well-typed. When desired, the stricter 'observation as measurement' from quantum mechanics can be modelled as a certain kind of observation in our sense: the change in behaviour associated with an 'observation as measurement' event is simply specified to conform to the uncertainty principle. The same holds for the constraint of quantum mechanics that only certain (i.e. commuting) observables may be measured simultaneously: whilst two events, say A and B, may be observed independently, their simultaneous observation constitutes a third event, AB say, with the different behavioural consequences dictated by quantum mechanics.

3.6 Decidable observation

In the theory of computation an observable is called *decidable*, or *effective*, if and only if its behaviour is given by a computable function. For example it is known to be undecidable whether or not a program terminates i.e., there is no algorithm for its determination [8].

We make no assumption about the decidability of an observable. The reason is simple. The field of Formal Methods within Computer Science [22] concerns itself with the mathematical specification and development of information systems. Typically a specification embodies a twofold constraint: the required program must conform to such-

and-such a functional specification *and terminate*. Without the last conjunct, undesirable programs which execute forever, never yielding a result, might be allowed (in some models of computation). But such a specification is no more than a behaviour phrased in terms of observables for input, output (appearing in the functional specification) and termination (appearing in the second conjunct). And we have just seen that termination cannot be assumed to be decidable.

The consequence of allowing an observable to be undecidable is that some ingenuity is required to prove that an implementation meets a specification phrased in terms of its observables: no program can possibly achieve that task in general.

4 The philosophy of LoA

The time has come to provide further conceptual clarification concerning the nature and consequences of the method of abstraction. In this section we relate the relevant work of Marr, Pylyshyn, Dennett and Davidson to the method of abstraction, and discuss the thorny issues of pluralism and realism.

4.1 Levels of organization and of explanation

Several important ways have been proposed for speaking of levels of *analysis* or description of a system. Two are representative:

1. *Levels of organization*, LoOs, support an ontological approach in which the system under analysis is supposed to have a hierarchical structure in itself, or *de re*, that is allegedly uncovered by its description and objectively formulated in some neutral observation language [47]. For example levels of communication, of decision processing [32] and of information flow can all be presented as specific instances of analysis in terms of LoOs.
   There is a twofold connection between LoOs and LoAs. If the hierarchical structure of the system itself is thought of as a GoA then to each constituent LoA is a corresponding LoO. Alternatively one can conceive of the analysis of the system, not the system itself, as being the object of study. Then the method of abstraction leads one to consider a GoA whose con-

stituent LoAs are the LoOs. Note that since the system under analysis may be an artefact, knowledge of its LoO may be available only in terms of knowledge of its specifications.

2. *Levels of explanation*, LoEs, support an epistemological approach common in cognitive and computer science [6]. Strictly speaking, the LoEs do not really pertain to the system or its model. They provide a way to distinguish between different epistemic approaches and goals, such as when we analyse an exam question from the students' or the teacher's perspectives, and the description of the functions of a technological artefact from the expert's or the layperson's point of view.
   An LoE is an important kind of LoA. It is pragmatic and makes no pretence at reflecting an ultimate description of the system. It has been defined with a specific practical view or use in mind. Manuals, pitched at the inexpert user, indicating 'how to' with no idea of 'why', provide examples.

The two kinds of 'structured analysis' are of course interrelated. Different LoEs (e.g. the end-user's LoE of how an applications package is to be used versus the programmer's LoE of how it is executed by machine) are connected with different LoAs (e.g. the end-user's LoA represented by a specific graphic interface versus the programmer's code), which in turn are connected with different LoO (e.g. the commonsensical WYSIWYG versus the software architecture). However LoAs provide a foundation for both, and LoOs, LoEs and LoAs should not be confused.

One of the most interesting examples of multi-layered analysis is provided by Marr's influential three-level hypothesis. After Marr [31], it has become common in cognitive and philosophical studies [30] to assume that a reasonably complex system can be understood only by distinguishing between levels of analysis:

> Almost never can a complex system of any kind be understood as a simple extrapolation from the properties of its elementary components. Consider for example, some gas in a bottle. A description of thermodynamic effects – temperature, pressure, density, and the relationships among these factors – is not formulated by using a large set of equations, one for each of the particles involved. Such effects are described at their own level, that of an enormous collection of particles; the effort is to show that in principle the microscopic and the macroscopic descriptions are consistent with one another. If one hopes to achieve a full understanding of a system as complicated as a nervous system, a developing embryo, a set of metabolic pathways, a bottle of gas, or even a

> large computer program, then one must be prepared to contemplate different kinds of explanation at different levels of description that are linked, at least in principle, into a cohesive whole, even if linking the levels in complete detail is impractical. For the specific case of a system that solves an information-processing problem, there are in addition the twin strands of process and representation, and both these ideas need some discussion. ([31] pp. 19-20).

More specifically, in the case of an information-processing system, Marr and his followers suggest the adoption of three levels of analysis:

1. the computational level. This is a description of 'the abstract computational theory of the device, in which the performance of the device is characterised as a mapping from one kind of information structures, the abstract properties of this mapping are defined precisely, and its appropriateness and adequacy for the task at hand are demonstrated' [31], p. 24;
2. the algorithmic level. This is a description of 'the choice of representation for the input and output and the algorithm to be used to transform one into the other' (p. 24-25);
3. the implementational level. This is a description of 'the details of how the algorithm and representation are realized physically – the detailed computer architecture, so to speak' (p. 25).

The three levels are supposed to be loosely connected and in a one-to-many mapping relation: for any computational description of a particular information-processing problem there may be several algorithms for solving that problem, and any algorithm may be implemented in several ways.

Along similar lines, Pylyshyn [40] has spoken of the semantic, the syntactic, and the physical levels of description of an information-processing system, with the (level of) functional architecture of the system playing the role of a bridge between Marr's algorithmic and implementational levels. And Dennett [11] has proposed a hierarchical model of explanation based on three different 'stances': the intentional stance, according to which the system is treated as if it were a rational, thinking agent attempting to carry on a particular task successfully; the design stance, which concerns the general principles governing the design of any system that might carry on those tasks successfully; and the physical stance, which considers how a system implementing the appropriate design-level principles might be physically constructed.

The tripartite approaches of Marr, Pylyshyn and Dennett are each readily formalised in terms of GoAs with three LoAs. It is important to observe that they are primarily concerned with, and so assign a privileged role to, explanations. As a result, their ontological commitment is embedded. The reasoning, were it to be made explicit, would be: 'this is the right level of analysis because that is the right LoO', where no justification is offered for why that LoO is the right one. Nor is the epistemological commitment made explicit or defended; it is merely presupposed. This is where the method of abstraction provides a significant advantage. By starting from a clear endorsement of each specific LoA, a strong and conscious effort can be made to uncover the ontological commitment of a theory (and hence of a set of explanations), which now needs an explicit acceptance on the part of the user, and requires no hidden epistemological commitment, which now can vary depending on goals and requirements.

4.2 Conceptual schemes

The resemblance between LoAs and conceptual schemes (CSs) is close enough to require clarification. In this section, we shall briefly compare the two. The aim is not to provide an exegetical interpretation or a philosophical criticism of Davidson's [10] famous attack on the possibility of irreducible CSs, but rather to clarify further the nature of LoAs and explain why LoAs can be irreducible, although in a sense different from that liked by supporters of the irreducibility of CSs.

According to Davidson, CSs share the following features:

1. CSs are clusters or networks of (possibly acquired) categories. 'Conceptual schemes, we are told, are ways of organizing experience; they are systems of categories that give form to the data of sensation; they are points of view from which individuals, cultures, or periods survey the passing scene' ([10], p. 183).
2. CSs describe or organise the world or its experience for communities of speakers. 'Conceptual schemes (languages) either organize something, or they fit it', and as 'for the entities that get organized, or which the scheme must fit, I think again we may detect two main ideas: either it is reality

(the universe, the world, nature), or it is experience (the passing show, surface irritations, sensory promptings, sense-data, the given)' ([10], p. 192).

3. CSs are inescapable, in the sense that communities of speakers are entrapped within their CSs.
4. CSs are not intertranslatable.

Davidson argues against the existence of CSs as inescapable (from within) and impenetrable (from without) ways of looking at the world by interpreting CSs linguistically and then by trying to show that feature 4. is untenable. Could the strategy be exported to contrast the existence of equally inescapable and impenetrable LoAs? Not quite.

Let us examine what happens to the four features above when LoAs are in question:

a. LoAs are clusters or networks of observables. Since they deal with observables, LoAs are not an anthropocentric prerogative but allow a more general (or indeed less biased) approach. We do not have to limit ourselves to human beings or to communities of speakers. Different sort of empirical or abstract agents – human beings but also computers, animals, plants, scientific theories, measurement instruments etc. – operate and deal with the world (or, better, with the data they glean from it) at some LoAs. By neatly decoupling LoAs from the agents that implement or use them, we avoid the confusion between CSs, the languages in which they are formulated or embodied, and the agents that use them. We shall return to this point presently.
b. LoAs model the world or its experience. LoAs are anchored to their data but they do not describe or organise them, they actually build models out of them. So the relation between models and their references (the analysed systems) is neither one of *discovery*, as in Davidson's CSs, nor one of *invention*, but one of *design*, to use some general categories. It follows that, contrary to Davidson's CSs, it makes no sense to speak of LoAs as Xerox machines or personal organisers of some commonly shared ontology (the world or its experience). Ontological commitments are initially negotiated through the choice and shaping of LoAs, which therefore cannot presuppose a metaphysical omniscience.

Because of the differences between 1.-2. and a.-b. the remaining two features acquire a significantly different meaning, when speaking of LoAs. Here is how the problem is reformulated. LoAs generate (and commit the agent to) information spaces. In saying that some LoAs can be irreducible and hence untranslatable we are not arguing that:

i. agents using LoAs can never move seamlessly from one information space to another. This is false. They obviously can, at least in some cases, just imagine gradually replacing some observables in the LoAs of an agent. This is equivalent to arguing that human being cannot learn different languages. Note, however, that some agents may have their LoAs hardwired, imagine, for example, a thermometer;
ii. agents using LoAs can never expand their information spaces. This is also false. Given the nested nature of some LoAs and the possibility of constructing supersets of sets of observables, agents can aggregate increasingly larger information spaces. This is equivalent to arguing that human speakers cannot expand their languages semantically, an obvious nonsense.

So, if we are talking about the agents using or implementing the LoAs, we know that agents can sometime modify, expand or replace their LoAs, hence some degree of intertranslability, understood as the acquisition or evolution of new LoAs, is guaranteed. The point in question is another, however, and concerns the relation between the LoAs themselves.

LoAs are the place at which (diverse) independent systems meet and act on or communicate with each other. If one reads carefully one will notice that this is the definition of an interface. The systems interfaced may adapt or evolve their interfaces or adopt other interfaces, as in i. and ii., yet different interfaces may still remain mutually untranslatable. Consider the 'tasting LoA' and the 'purchasing LoA' in our wine example. But if two LoAs are untranslatable, it becomes perfectly reasonable to assume that

iii. agents can inhabit only some types of information spaces in principle.

Some information spaces may remain not just in practice but in principle inaccessible, or accessible only asymmetrically, to some agents. Not only, but given the variety of agents, what is accessible to one or some may not be accessible to all. This is easily explained in terms of modal logic and possible worlds understood as information spaces. The information space of a child is asymmetrically accessible from the information space of an adult, and the information space of a bat is only insufficiently overlapping with the information space of any human agent to guarantee a decent degree of translatability [36].

In principle, some information spaces may remain forever disjoint from any other information spaces some agents may be able to inhabit. When universalised, this is Kant's view of the noumenal world, which is accessible only to its creator. Does this imply that, after all, we are able to say what a radically inaccessible information space would be like, thus contradicting ourselves? Of course not. We are only pointing in the direction of the ineffable, without grasping it. It is a bit like drawing a figure without ever being able to paint it.

To return to Davidson, even conceding that he may be successful in criticising the concept of CSs, his arguments do not affect LoAs. The problem is that Davidson limits his considerations to information spaces that he assumes to be already linguistically and ontologically delimited, and when this is the case, one may concede his point. However, LoAs do not vouch for the kind of epistemic realism, verificationism, panlinguism and representationist view of knowledge that Davidson implicitly assumes in analysing CSs. And once these fundamental assumptions are eliminated, Davidson's argument loses its strength. Incommensurable and untranslatable LoAs are perfectly possible. We have also seen that this provides no good ground for a defence of some form of radical conceptual relativism. Davidson's criticism ends by shaping an optimistic approach to the problem of the incommensurability of scientific theories that we cannot share.

What conclusions can be drawn, from our analysis of LoAs, about the anti-realist reading of the history of science? An unqualified answer would fall victim to the same fallacy of un-layered abstraction we have been denouncing in the previous pages. The unexciting truth is that different episodes in the history of science are more or less comparable depending on the LoA adopted. Consider the great variety of building materials, requirements, conditions, needs and so on that determine the actual features of a building. Does it make sense to compare a ranch house, a colonial home, a town house, a detached house, a semidetached house, a terraced house, a row house, a cottage, a thatched cottage, a country cottage, a flat in a single-storey building, and a Tuscan villa? The question cannot be sensibly answered unless one specifies the LoA at which the comparison is to be conducted. Likewise, our answer concerning the reading of the history of science is: given the nature of LoAs, it is always possible to formulate a LoA at which comparing different episodes in the history of science makes perfect sense. But do not ask absolute questions, for they just raise absolute mess.

4.3 Pluralism

A LoA qualifies the level at which a system is considered. In this paper, we argue that it must be made precise before the properties of the system can sensibly be discussed. In general, it seems that many disagreements might be clarified and resolved by the various 'parties' by making precise their LoA. Yet another crucial clarification is now in order. It must be stressed that a clear indication of the LoA at which a system is being analysed allows pluralism without falling into relativism or perspectivism [19]. As remarked above, it would be a mistake to think that 'anything goes' as long as one makes explicit the LoA, because LoAs can be mutually comparable and assessable, in terms of inter-LoA coherence, of their capacity to take full advantages of the same 'constraining affordances' and of their degree of fulfilment of the explanatory and predictive requirements laid down by the level of explanation. Thus, introducing an explicit reference to the LoA clarifies that the model of a system is a function of the available observables, and that it is reasonable to rank different LoAs and to compare and assess the corresponding models.

4.4 Realism

For a typed variable to be an observable it must be interpreted, a correspondence that has inevitably been left informal. That interpretation cannot be omitted: a LoA composed of typed variables called simply x, y, z and so on and treated rather formally, in mathematics, would leave the reader (or the writer some time later) with no hint of its domain of application. Whilst that is the benefit of mathematics, enabling its results to be applied whenever its axioms hold, in the method of abstraction it confers only obscurity.

But does the informality of such interpretation hint at circularity or infinite regress? Must the system under consideration already be observed before an observation can be defined? The mathematics underlying our definitions of typed variable and behaviour has been indicated (even if it is not always fully used in practice) to make the point that in principle the ingredients in a LoA can be formalised. Thus there can be no circularity: the heuristically appreciated system being modelled never exists on the same plane as that being studied methodically.

The point might be made analogically by consideration of Tarski's well-known model-theoretic definition of truth [50]. Is there circularity or regress there? Might it be argued that one needs to know truth before defining it? The same resolution is to be offered here. Tarski's recursive definition of truth over syntactic construction is based on an appreciation of the properties truth is deemed to have, but that appreciation and the rigorous definition exist on 'different planes'. So circularity is avoided. More interesting is the question of infinite regress. Tarski's definition formalises certain specific properties of truth; a regress would obtain only were a complete characterisation sought. So it is with the interpretation required to define an observable. Some property of an undisclosed system is being posited at a certain level of abstraction. An unending sequence of LoAs could possibly obtain were a complete characterisation of a system sought. The primary candidate seems to be the mind, and that brings us back to the fundamental question, raised in the Introduction, of whether some systems require indefinite approximation. But that is not regress.

It is implicit in the method of abstraction that a GoA is to be chosen that is accurate or 'realistic'. How, then, is that to be determined without circularity? We give the answer traditionally offered in Mathematics and Science: it is determined by external adequacy and internal coherence. The behaviours at a moderated LoA must adequately reflect the phenomena sought; if not then either the definition of the behaviour is wrong or the choice of observables is inappropriate. When the definition of observables must incorporate some 'data' the latter may behave like constraining affordances and so limit the possible models. We refer to [14, 52] for further details and examples. Secondly, the condition embodied in the definition of GoA is a remarkably strong one, and ensures a strong degree of internal consistency or coherence between the constituent LoAs. The multiple LoAs of a GoA can be thought of as interlocking like the answers to a cross word puzzle. Though such consistency does not guarantee that your answer to the cross-word is the same as the originators, it drastically limits the number of solutions making each more likely.

Coherence and adequacy neither entail nor support naive realism. In one view, GoAs ultimately *construct* models of systems [13]. They do not represent or portray or uncover the intrinsic nature of the systems they analyse. By that view we understand systems derivatively, only insofar as we understand their models. Then coherence and adequacy are the most we can hope for.

5 Conclusion and further work

In this paper, we have introduced the method of abstraction and applied it to the study, modelling and analysis of phenomenological and conceptual systems. We have demonstrated its principal features and main advantages. The method clarifies implicit assumptions, facilitates comparisons, enhances rigour and hence promotes the resolution of possible conceptual confusions. We have argued that, for discrete systems, whose observables take on only finitely-many values, the method is indispensable. Its limitations, on the other hand, are those of any typed theory. Use of LoAs is effective in precisely those situations where a typed theory would be effective (where it is possible or desirable). Introduction of LoAs is often an important step prior to mathematical modelling of the phenomenon under consideration. But even when that further step is not taken, introduction of LoA remains a crucial tool in conceptual analysis. Of course, care must be exercised in type-free systems, where the use of LoA may be problematic. Such systems are susceptible to the usual paradoxes and hence to inconsistency [43, 23] when formalised mathematically but also when considered informally. Examples of such systems arise frequently in Philosophy and in Artificial Intelligence. However, we have shown that, if carefully applied, the method confers remarkable advantages in terms of consistency and clarity.

Of course the adoption of the method of abstraction raises important further questions. By way of conclusion, we call the reader's attention to five that seem particularly pressing:

1. Why do certain LoAs, e.g. the so-called 'naive physics' view of the world, appear 'privileged'?
2. What are the relations between GoAs [38]?
3. To what degree might a LoA determine semantics?
4. What kind of ontological commitment is entailed by (the acceptance of) a LoA?
5. How do LoAs account for the building blocks of information [52]?

Bibliography

[1] D. Adams, *The Hitchhiker's Guide to the Galaxy,* Pan Books Ltd., London 1979.

[2] M.A. Arbib, *The Metaphorical Brain II: Neural Networks and Beyond*, John Wiley & Sons, New York 1989.

[3] J. Barwise and J. Echmenedy, *The Liar: an essay on truth and circularity*, Oxford University Press, Oxford 1987.

[4] W. Bechtel and R.C. Richardson, *Discovering Complexity: Decomposition and Localization as Strategies in Scientific Research*, Princeton University Press, Princeton (NJ) 1993.

[5] J.S. Bell, 'The theory of local beables', in *Speakable and Unspeakable in Quantum Mechanics*, (collected papers on quantum philosophy), Cambridge University Press, Cambridge 1993, pp. 52-62.

[6] P. Benjamin, M. Erraguntla, D. Delen and R. Mayer, 'Simulation modeling and multiple levels of abstraction', in D.J. Medeiros, E.F. Watson, J.S. Carson and M.S. Manivannan (eds.), *Proceedings of the 1998 Winter Simulation Conference*, Washington (DC) 1998, pp. 391-398.

[7] M.A. Boden (ed.), *The Philosophy of Artificial Life*, Oxford readings in philosophy, Oxford University Press, Oxford 1996.

[8] G.S. Boolos, J.P. Burgess and R.C. Jeffrey, *Computability and Logic*, 4th ed., Cambridge University Press, Cambridge 2002.
[9] P. Cariani, 'Emergence and artificial life', in [28], 1991, pp. 775-797.
[10] D. Davidson, 'On the Very Idea of a Conceptual Scheme', originally published in *Proceedings and Addresses of the American Philosophical Association*, 47, 1974, reprinted in *Inquiries into Truth and Representation*, Clarendon Press, Oxford 1984, pp. 183-98. All page numbers to the quotations in the text refer to the reprinted version.
[11] D.C. Dennett, *The Intentional Stance*, MIT Press, Cambridge (MA) 1987.
[12] A. Egyed and N. Medvidovic, 'A formal approach to heterogeneous software modeling', in *Lecture Notes in Computer Science*, Springer Verlag, 1783, p. 178, 2000.
[13] S. Fish, *Is There a Text in This Class?*, Harvard University Press, Cambridge Massachusetts 1980.
[14] L. Floridi, 'Information', in L. Floridi (ed.), *The Blackwell Guide to the Philosophy of Computing and Information*, Blackwell, Oxford, New York 2003.
[15] L. Floridi and J.W. Sanders, 'On the morality of artificial agents', CEPE 2001, Computer Ethics: Philosophical Enquiry (Lancaster University, 14-16 December, 2001), forthcoming in A. Marturano and L. Introna (eds.), *Ethics of Virtualities. Essays on the limits of the bio-power technologies*. To be published in the series Culture Machine, Athlone Press, London. Preprint available at http://www.wolfson.ox.ac.uk/floridi/.
[16] L. Floridi and J.W. Sanders. *Artificial Life Observed*, in preparation.
[17] C. Geertz, *The Interpretation of Culturesi*, Basic Books Classics, New York 1973.
[18] M. Gell-Mann, *The Quark and the Jaguar*, Freeman, New York 1994.
[19] S.D. Hales and R. Welshon, *Nietzsche's Perspectivism*, University of Illinois Press, Urbana 2000.
[20] I.J. Hayes, *Specification Case Studies*, Prentice-Hall International, Hertfordshire 1986.

[21] H. Hendriks-Jansen, 'In praise of interactive emergence: or why explanations don't have to wait for implementations', in [27], 1989, pp. 282-299.
[22] C.A.R. Hoare and H. Jifeng, *Unifying Theories of Programming*, Prentice-Hall International, Hertfordshire 1998.
[23] D. Hofstadter, *Gödel, Escher, Bach*, Harvester Press, London 1979.
[24] P. Hughes and G. Brecht, *Vicious Circles and Infinity: A Panoply of Paradoxes*, Doubleday, Garden City (NY) 1975.
[25] J.A.S. Kelso, *Dynamic Patterns*, MIT Press, Cambridge (MA) 1995.
[26] D.E. Knuth, *The Art of Computer Programming, vol. 2, Semi-numerical algorithms*, Second edition, Addison Wesley, Reading (Mass.) 1981.
[27] C.G. Langton (ed.), *Artificial Life: Proceedings of an Interdisciplinary Workshop on the Synthesis and Simulation of Living Systems*, Santa Fe Institute Studies in the Sciences of Complexity, Proceedings 6, Addison-Wesley, Redwood City (CA) 1989.
[28] C.G. Langton, C. Taylor, J.D. Farmer and S. Rasmussen (eds.) *Artificial Life II*, Santa Fe Institute Studies in the Sciences of Complexity, Proceedings 10, Addison-Wesley, Redwood City (CA) 1992.
[29] C.G. Langton, *Artificial Life*, Reprinted in [7], 1996, pp. 39-94, updated version of the original which appeared in [28].
[30] R. McClamrock, 'Marr's three levels: a re-evaluation', *Minds and Machines*, 1, 1991, pp. 185-196.
[31] D. Marr, *Vision*, Freeman, New York 1982.
[32] M.D. Mesarovic, D. Macko and Y. Takahara, 'Theory of hierarchical, multilevel, systems', in R. Bellmann (ed.), *Mathematics in science and engineering*, Academic Press, New York 1970.
[33] T.M. Mitchell, *Machine Learning*. McGraw Hill, New York 1997.
[34] *Minds and Machines*, 11(1), 'The Turing test: past, present and future', guest editor J.H. Moore, February 2001.
[35] J.D. Murray, *Mathematical Biology*, Springer Verlag, Berlin, Heidelberg 1989.

[36] T. Nagel, 'What is it like to be a bat?', in *The Philosophical Review*, 83(4), 1974, pp. 435-450.

[37] A. Newell, *Unified theories of cognition*, Harvard University Press, Cambridge (MA) 1990.

[38] G. Paronitti, *Structure of abstraction*, in preparation.

[39] C.S. Peirce, 'Description of a notation for the logic of relatives, resulting from an amplification of the conceptions of Boole's calculus of logic', in *Memoirs American Academy*, 9, 1870, pp. 317-378.

[40] Z.W. Pylyshyn, *Computation and Cognition*, MIT Press, Cambridge (MA) 1984.

[41] J. Robinson (ed.), *The Oxford Companion to Wine*, Oxford University Press, Oxford 1994.

[42] J. Robinson, *Vintage Timecharts*, Weidenfeld and Nicolson, New York 1989.

[43] B. Russell, 'Letter to Frege', 1902, in J. van Heijenoort (ed.) *From Frege to Gödel: A Source Book in Mathematical Logic, 1879-1931*, Harvard University Press, Cambridge (MA) 1967, pp. 124-125.

[44] S.N. Salthe, *Evolving Hierarchical Systems*, Colombia University Press, New York 1985.

[45] F. de Saussure, *Cours de linguistique générale*, ed. C. Bally and A. Sechehaye, with the collaboration of A. Riedlinger, Payot, Lausanne and Paris, Engl. transl. W. Baskin, *Course in General Linguistics*, Fontana/Collins, Glasgow 1977.

[46] F. de Saussure, *Saussure's Third Course of Lectures in General Linguistics*, 1910-1911, 'From the Notebooks of Emile Constantin', *Language and Communication series*, vol. 12, trans. and ed. E. Komatsu and R. Harris, Pergamon, Oxford 1993.

[47] H.A. Simon, *The Sciences of the Artificial*, third ed., MIT Press, Cambridge (MA) 1996.

[48] J.M. Spivey, *The Z Notation: A reference manual, second edition*, Prentice-Hall International, Hertfordshire 1992.

[49] G. Stein, *Sacred Emily*, 1913, in *Selected Writings of Gertrude Stein*, Vintage books, New York 1990.

[50] A. Tarski, 'The Semantic Conception of Truth and the Foundations of Semantics', in *Philosophy and Phenomenological Re-*

search, 4, 1944, pp. 341-76, reprinted in L. Linsky (ed.) *Semantics and the Philosophy of Language*, University of Illinois Press, Urbana (IL) 1952.
[51] D'A.W. Thompson, *On Growth and Form*, the complete revised edition, Dover 1992, (reprint of the CUP 1942 edition).
[52] M. Turilli, *Perceptual information*, in preparation.
[53] A.M. Turing, 'Computing machinery and intelligence', in *Mind*, 59, 1950, pp. 433-460.

Luciano Floridi, Dipartimento di Scienze Filosofiche, Università di Bari – Faculty of Philosophy, Sub-Faculty of Computation Information Ethics Group, University of Oxford
Jeff W. Sanders, Programming Research Group, Oxford University Computing Laboratory Information Ethics Group, University of Oxford

Giuseppe Lanzavecchia

Models in science are unnecessary but useful

Abstract:

Humans use models in all their activities, those of scientists are conceptual models, and all the models are artificial objects. Models are necessary to give a 'perceivable' idea of science, and useful for forming conceptions of it and for orienting and predicting its developments. Some people think models are indispensable for actually doing science, at least in some cases, but this paper tries to show that this is not so, and that science proceeds by improving its results, and rejects models that are no longer useful; in other words, it 'falsifies' them. This way of operating poses problems for philosophy, and involves the science of the artificial in relation to the way humans are and how their minds work.

Keywords:
model, science, mathematical model, artificial, philosophy, falsification, solar system, gravity, relativity, quantum mechanics

1 Humans and models

For as long as the history of our species has been documented, but almost certainly ever since humans first appeared on the planet, they have used models in all their activities, especially to explain nature and its events – in other words, to do science.

Indeed, nearly all scholars (in the first instance scientists and philosophers) believe there can be no science without models, and many even think of science as a collection of the models that try to describe it. As a result, models are taken as references for all of the characteristics of science, starting from its value (or lack of value) as 'truth', and for understanding all its other properties, starting from its ethical value.

I happen to be one of the few people who think, instead, that science depends only on the method used to do it, which (in the last analysis) is still the same one described by Galileo (Lanzavecchia and Negrotti, 144-9). In a nutshell, it assumes that science starts with experimental observation of natural events, of whatever kind (physical or psychic) they may be; catalogues them; identifies their regularities, ignoring the aspects that seem nonessential; finds the rules that govern them and if possible translates them into mathematical equations; then verifies the rules over and over again. A model, then, of whatever kind, is a human construct, and observed properties do not match it in a double bivalent way. Most important, standards of truthfulness, falsity, ethics, and so on apply only to the model. Though the model may eventually turn out to be inadequate, the experimental knowledge on which it was based may still be true, in the sense that repeated trials always give the same result.

Scientific literature contains a good number of studies on the role, meaning and importance of models. I would mention, for instance, the proceedings of the conference on 'The Role of Models in Science and Knowledge' organized by Italy's Accademia dei Lincei in 1999 (Accademia Nazionale dei Lincei), in particular Carlo Cellucci's paper on 'Models, Analogies and Metaphors' (Cellucci). Analogy and metaphor play an important role in models, bringing into play the typical devices of human thought: intuition, correlation, logical analysis, reasoning, construction of a frame of reference. All of them are useful for humans, but to my mind they are altogether foreign to the profound reality of nature, hence to the scientific method (Lanzavecchia and Negrotti, 91-100).

In a word, a model is an artificial construct that seeks to portray reality, hence to substitute for science. Like any other artificial construct, however, it can do no more than reproduce one or at best a few partial characteristics (if indeed it manages to do that), while it adds others that have nothing, or practically nothing, to do with reality (Negrotti; Lanzavecchia and Negrotti, 85-91).

The ultimate question is this: Can there be science, hence scientific theories, that do not use models? We shall see that it is indeed possible, because actual examples exist, though they are a very small minority compared with the usual practice.

2 The usefulness and power of models

All humans use models, because that is our natural way to address problems, though scientists know – or should know – that their use is not part of the scientific method.

Models are often indispensable. We would not know how to handle data from observations of natural events without formulating hypotheses as to how and why they happen; how they relate to each other, to events produced by other groups of phenomena and to models made of them; and the mechanisms that govern them. Scientists have intuitions, conceive hypotheses and build models. Often such models are mathematical; the better a model can account for natural events, the more mathematical it is likely to be.

Besides providing a frame of reference and an 'explanation' of scientific phenomena, models have a heuristic property. When we say one model is 'better' than another, we mean it is better at predicting a series of other phenomena that go beyond the ones used to build the model. This property, which makes it possible to 'build' results – which must in any case be rigorously validated through experimentation – makes modeling a particularly attractive activity.

In fact, building models comes down to the same thing as building and organizing concepts, or categorizing, one of the most important of all the creative and thought-structuring processes (Murphy; Smith and Medin) that make up our 'natural' intuitive way of doing things. Nonetheless, science is anti-intuitive, and in this respect should be considered one of humanity's 'heretical' endeavors (Lanzavecchia and Negrotti, 121-143, in particular 128-131).

This 'power' that models have – the fact that they are nearly always used in doing science and originate nearly all theories, the conviction that a growing number of phenomena require the use of models if they are to be studied scientifically, because they involve aspects beyond the reach of direct experimentation, whether for spatial reasons (as in studies of the sun and stars, black holes, or the center of the earth) or temporal reasons (as in studies of past cultures and societies, or of cosmology) – contributes decisively to confusion between science and models. Philosophers and epistemologists are especially prone to this misunderstanding, and so are a good number of scientists.

In this way, science – which intends to describe nature – becomes instead an artificial construct; as a result, judgments and values refer not to data discovered by science, but to the construct that is supposed to represent them.

3 Are there cases in which science must use models?

Among the cases mentioned above, let us consider that of the structure, composition and properties of the sun, as discussed in 'The Sun and Solar Wind: A Search for the Beginning' (The Sun).

This report begins with the observation that children – and scientists too – often play with models that are not always necessary. But if objects are too small to be observed directly, or if they are inaccessible, or if (as in the case of gravity, magnetism and energy) they are invisible and can be studied only indirectly, through their effects on matter, then in one way or another they have to be modeled.

The report then points out that some models are physical (globes and mechanical models of the solar system, for instance), but others are mental images (e.g., Bohr's solar model of the atom) and still others are mathematical. Models are perfected through the acquisition of new, more exact scientific knowledge.

Since it is impossible to perform direct experiments on the sun, scientists have made models of it. One is the Standard Solar Model (Bahcall), which uses certain data (spherical shape, diameter, mass, luminosity, approximate age) and assumes certain hypotheses:

(1) the sun originated from a primordial cloud of hydrogen mixed with a small amount of helium;
(2) it is in a stationary state;

(3) the hydrogen in its core undergoes a process of nuclear fusion that generates helium and energy;
(4) the energy is conveyed outward by radiation to the opaque zone, where convection dominates. The temperature at the center of the core is calculated to be 15.6 mega°K and its density is put at 148,000 kilograms per cubic meter.

How can these hypotheses and calculated values – that is, the model – be verified? Its level of trustworthiness depends on the quality of the data and values that it can provide on observable quantities, including (1) the production of neutrinos and their detection on earth, and (2) seismic activity in the sun, as measured through the oscillations of the solar photosphere.

We can conclude easily enough (this is not a textbook on solar physics) that this model seems to match better and better – via continuous feed-in and feed-back of experiments, data, hypotheses, verifications and adjustments – the observable and measurable signals that the hypothetical structure of the sun is supposed to provide.

The same sort of reasoning, though more complex, applies to cosmology, the study of the birth of the universe, with the added complication that verifications made today refer to events that occurred billions of years ago.

4 How science accesses the inaccessible

Science started from sensations, with Plato and above all Aristotle, and over the centuries the analysis of sensations was sharpened through the work of philosophers and scientists (physicists and psychologists), and eventually Husserl's phenomenology (Husserl 1913). Ernst Mach, looking into the way senses and sensations are translated into science, concluded that scientific concepts and theories do not reflect the nature of things; rather, they are pragmatic tools used to organize sensorial data in an economic way – i.e., effective for the purposes of prediction (Mach 1906).

This pragmatic approach is precisely what enables us to prune 'metaphysical' meanings from the role of the senses. In medieval times, for instance, many people thought that eyeglasses 'deformed' reality and so should not be used to investigate nature and do science. The same was said of Galileo's telescope and his observations of Jupiter's moons. Later on, as a better understanding was gained of how our senses work (sight for light, hearing for sound, and so on), people came to accept contributions from 'visions' and 'sensations' gained with infrared and ultraviolet rays or ultrasound, and to accept any 'sensorial' aspect, in the broad sense, that can offer measurements of physical (and psychic) properties, for instance optical, mechanical, magnetic and electrical. Any measurable property thus became a means (much like a 'sense') for studying nature.

Remote effects – those caused in one place and effectuated in another, without intermediaries – were more difficult to accept. Galileo did not, and in the *Dialogues* he gave an ingenious but false explanation of tides. Newton observed that interaction among bodies, even at a distance (the apple falling to the earth, the earth and the sun, Jupiter and its moons) occurs as if there existed a force whose characteristics are always the same (Newton). Today we can measure this force, therefore it exists.

It has been still more difficult for people to accept the idea that a measurement made today can explain (or, rather, describe) something that happened billions of years ago. But in the meantime, scientists had to address the issue of what simultaneous events are, and how to measure events that last only a few fractions of a billionth of a second, when what we perceive as an instant is little less than a tenth of a second. The relationships between human time and the infinitesimal time of submicroscopic events or the very long time of cosmic events are more or less similar, and if we wish to describe such phenomena, we are forced to accept these relationships as sensorial experiences.

If we have measurements – 'sensations' measured with suitable instruments – that are much more precise and reproducible than those we can obtain with our five inborn senses, we can work our way backward along the same path, starting from the model used to reach the verified results, remove unnecessary hypotheses from theories, and return to the experimental premises and verified conclusions. In this way, we can eliminate models (in other words, do without the excessively artificial) and simply do science.

5 Science retains and improves its experiences, falsifies and rejects models that are no longer useful

To validate the above statement, I shall consider only one case, but an especially significant and widely discussed one: the solar system as viewed in the heliocentric and geocentric hypotheses, in Newton's and in Einstein's.

Our starting point is Ptolemy's description of the solar system in the *Almagest* (2nd cent. A.D.), where he shows the earth at the center and the planets and the sun rotating around it. This is a picture of the solar system as seen by an observer on earth, and when it first appeared, well before Ptolemy, it was most likely understood as a mathematical description of the motion of the planets, one that made it possible to see how their positions change over time (Russo).

Conversely, Nicolaus Copernicus' considerations and the observations made by Tycho Brahe and Johannes Kepler, which led to Newton's model of gravity, describe the solar system as seen by an observer located on the sun. Nonetheless, both descriptions represent the same movements of the heavenly bodies, and can therefore be translated the one into the other via geometric transformations of the reference systems. Ptolemy's and Copernicus' orbits are circular, Kepler's are elliptic. This difference may have been due to more accurate measurements, since Kepler's ellipses are very nearly circular. The Greeks knew about ellipses; Apollonius of Perga developed a whole rigorous theory of conic sections.

From the mathematical standpoint, the description of the heliocentric system was much simpler than that of the geocentric system, where the motion of the planets was a combination between a uniform circular orbit around the earth (deferent) and other orbits (epicycles) centering on deferents of their own. It was mainly this use of geometry that led Copernicus to formulate the heliocentric theory.

As Archimedes noted in his treatise *On the Sphere and the Cylinder*, Aristarchus had proposed a heliocentric theory as far back as the third century B.C. Other Greek astronomers and mathematicians too believed that the earth revolved around the sun, and as early as the fourth century B.C. Heraclides of Pontos held that the earth revolved around its own axis once a day.

Each of the different 'models' of the solar system requires an analysis of its own. Some mathematical models simply describe the movements of the planets; the heliocentric one is handy, the geocentric one reflects what we see from the earth. But the choice between earth and sun involves philosophical and religious concepts, not scientific ones. For the Greeks, the heavenly bodies moved along 'tracks'; for Kepler, they were guided by angels.

Newton, with his formula F=MA, explained not only the movements and laws that Kepler had found, but also the interaction among bodies of any size, wherever they may be (on our planet or in the farthest reaches of space), in the past, the present and the future. He made it clear, however, that though all this happens as if there were some force having very specific universal characteristics, he was not going to try to explain what that force is. In other words, Newton's model was simply the exact mathematical description of what the astronomers of his day had observed. All the rest, for science, could not – and cannot – be other than stabs in the dark.

Newton's mechanics is still valid. Engineers still use it in all their constructions, though Einstein's relativistic mechanics is necessary to describe very large systems and very high speeds. But the structure of the formulas of mechanics has not been knocked out of kilter; they have merely been refined (e.g., the composition of speed, the value of mass).

What has been revolutionized is the concept of space and time that Newton – and Kant too, who translated it into philosophical terms – had adopted on the basis of intuitive considerations and pure reasoning, but not scientific observations. It would be too time-consuming and complex to go into the matter here; it has been treated by many authors, and I intend to discuss it in another essay (Lanzavecchia 2003). In a nutshell, Newton introduced the concept of *absolute space* (Leibniz denied its objective and absolute reality), and Kant thought of space and time as preceding experience, being part of humans' subjective makeup. For Kant, space was a given of sensitive intuition, a structure that transcends all experience and cannot be subjected to experimentation, as can all other events. In this sense, I think, space in

Kant's conception is a category, a way of knowing reality, a given of intellect that determines the transcendental conditions of experience.

Unfortunately, these concepts of mind, thought and reality, lacking the necessary experimental bases, are not only unscientific; if applied to science, shaping its models, they become a source of error. It is safe to say that practically all of Kant's categories are unacceptable because they clash with experience or do not stand up under Gödel's logical analysis (Gödel). Quantum mechanics alone has shot a good number of them full of holes, starting with causality.

In the 19th century the physicist Heinrich Rudolph Hertz (Hertz; Sant'Anna) worked out a theory of mechanics that does not use the concept of force; instead, he translated forces into constraints. When I was a student, I tried to use Hertz's mechanics to describe the motion of a planet around the sun, attributing to the space in which the planet moves an appropriate 'metrics' that translates Newton's or Einstein's forces into constraints. Unaffected by forces, the planet's motion must be rectilinear and uniform, but because of the metrics that 'deforms' space, we perceive that motion – translated into normal Euclidean space – as describing an ellipse.

I think this example shows how a model based on concepts not derived from experience leads to material errors or incorrect interpretations, even if the mathematical equations derived from such a model can make correct predictions. In particular, the concepts of space and time should be calibrated on the world that is being observed: the space-time of relativity, the space of the phases of statistical mechanics, or the particle space of quantum mechanics: spaces and times that involve aspects like the simultaneity, identity and the very nature of the thing observed (whether, for instance, it is a wave or a corpuscle). I would add that when I was young, the space in relativity theory was thought to be 'curved' and closed in on itself; today, our knowledge of cosmology leads us to think that it is essentially 'flat', and deformed only around material masses.

In this section we have seen how our scientific knowledge of the solar system has been refined and enriched, while 'ideological' incrustations – often not even noticeable, and at any rate foreign to the scientific method – have been gradually chipped away.

6 Can science do without models?

Freed of their ideological superstructures, the Ptolemaic and Copernican systems are examples of physics theories that are not based on any model, but simply describe in mathematical terms the motions of the heavenly bodies as seen from a given viewpoint, and make it possible to predict all their future movements.

However, if we wish to explain the mechanisms that determine these movements, the pre-Newtonian models are unhelpful and Newton's, as improved by Einstein, is correct. What is most interesting is that scientists have been finding more and more experimental corroboration for the theory of gravity – measurements of gravity, of gravity waves and of their speed, which is practically the same as the speed of light (Brumfiel) – and of all the characteristics implied by the theory.

The more all the aspects of a theory are corroborated by experimental evidence, the more that theory can – and should – do without modeling hypotheses. Today this applies to all the most important scientific theories and, as Alan Cromer has remarked, for the first time in the history of mankind, we possess real knowledge of the nature of existence and of our place in it. Cromer adds that Academics are alarmed when they hear talk of 'truth' and 'certainty'. They believe that truth and certainty are not possible, because philosophers have shown that there can be no knowledge, either empirical or deductive, free of error. But in the case of a finite number of discrete entities, such as the chemical elements or human genes, 'certainty' is an appropriate word. And in any case, our knowledge of atoms and genes is just as certain as our knowledge of tables and chairs (Cromer) – regardless of any model.

The most striking example of a theory that does without a model is quantum mechanics, despite the fact that it is nearly always presented with and symbolized by Schrödinger's equation, the casual result of a model of the electron that seeks to justify its dual properties as a wave and a corpuscle.

Erwin Schrödinger's mechanics is often called 'ondulatory', while 'quantum mechanics' is the term used for Werner Heisenberg's, but the two have been shown to be equivalent.

Quantum mechanics is based on the concept of a quantum operator, a mathematical entity that corresponds to certain classical quantities (e.g., energy) and assumes only discrete values. The operator is not a function; it represents the operations that need to be performed on a function in order to describe quantum systems, to transform the quantities of classical physics into quantum quantities, to correlate quantum quantities and entities to each other and, starting from measurements of quantum phenomena, to predict their results.

However, only measurable quantities and phenomena have meaning in quantum mechanics. The value of a physical quantity exists only after it has been measured, and establishing an apriori value is meaningless. Hence it makes no sense to speak of corpuscles or waves or orbits. Heisenberg used to rage against the idea of electron orbits, asserting that they have never been observed and cannot be observed, therefore they do not exist.

The same point can be made about the quantum mechanics developed by Paul André Marie Dirac, anti-modelist par excellence. Dirac formulated his equations in a way that took account of all the experimental events – including relativity, spin and Pauli's exclusion principle – that had been ignored in previous theories. His equations describe atoms and subatomic particles perfectly, and thanks to the enormous and ever-growing power of computers, they are now widely used despite the laborious calculations they require.

7 Philosophy and scientific theories

Quantum mechanics posed new questions for scientists and philosophers regarding the nature of reality, the probabilistic interpretation of quantities and events, discrepancies between our observations of the physical world and the behavior of matter on the one hand and the logical structure of human thought on the other, and many other aspects as well: space-time, causality, relationships among quantities of which some cannot be exactly measured simultaneously, symmetries (matter and antimatter, for instance), and transfers of matter and properties.

Einstein refused to accept the hypothesis that 'God plays dice'. Many physicists, including Louis de Broglie*, have tried to figure out what 'hidden' (hence untestable) parameters give quantum mechanics the same features of rationality that characterize human thought and classical science. John Bell (Bell) has described 'inequalities' that can be used to verify whether quantum mechanics is or is not correct. To date quantum mechanics has always proved right. No other scientific theory gives answers as precise as quantum mechanics does, or makes it possible to set up equally precise experiments, notwithstanding the recurrent question of indetermination.

Quantum mechanics is perfectly determined; what are called 'indeterminations' are a characteristic of matter, not something that makes it impossible for science to do certain things (like trying to pour a gallon of water into a quart bottle; it just can't be done). Statistical interpretation is none other than the verifiable fact that the square of Shrödinger's ψ wave function for a set of particles represents the probability of finding them in the places (or moving at the speeds, or with the energies) indicated by the variables on which ψ depends (Born). Statistical methods were used in classical physics – in thermodynamics, for example – well before the advent of quantum mechanics. By the same token, phenomena in our everyday lives are probabilistic. It is most unlikely that a brick will ascend on its own to the waiting workman on the third floor, yet the probability is not absolutely nil; we can entertain the idea that the molecules in the air might for some reason strike the bottom of the brick instead of the top.

Galileo and Newton were both concerned to avoid any metaphysical speculation ('hypotheses non fingo' said Newton), but later on, when the laws of mechanics were known perfectly, some scientists began to try to trace them back to enunciated principles, thus suggesting that their origin is non-empirical; in other words, they created

* In 1954 I had the good fortune to be invited to dinner at Louis de Broglie's home in Paris, together with and thanks to my old teacher Professor Giovanni Polvani. There was a lively discussion on the possibility of eliminating what de Broglie, Einstein and many others took to be the unacceptable aspects of quantum mechanics, by introducing 'paramètres cachés' that operate mathematically but cannot be verified experimentally. This would introduce a reality extraneous to human perception, though in agreement with the logical structure of our minds.

models, some of them very intelligent. The most famous one was Pierre-Louis Maupertuis' 'law of the least quantity of action' which he derived from the teleological idea that nature, like man, operates with a purpose and seeks to attain it with the minimum expense of action (energy).

But what do philosophers have to say about science, theory and models? First of all, I would cite Ernst Mach, who contributed more than anyone else to laying the foundations of anti-metaphysical and essentially scientific thought, based on a radical empiricism that was critical of preconceived references – such as the Newtonian concepts of absolute space, time and speed – for which there is no experimental evidence (Mach 1883). For H. Reichenbach, who tried to translate relativity and quantum mechanics into philosophical terms, a scientific theory is no more than a formal system that requires, however, a physical interpretation filtered through coordinating definitions (Reichenbach 1944, 1951 and 1958), though according to the logical positivists such definitions cannot be verified and are therefore meaningless.

The great majority of philosophers have been concerned not so much with science as with how scientists do it. Paul K. Feyerabend saw 'methodological anarchism' as the best tool to advance scientific knowledge. Lakatos (Imre Lipsitz) applied his teacher Karl Popper's idea of falsifiability to mathematics, and to do so had to make mathematics empirical – even the kind of mathematics that is not science but an intellectual construct. Popper was struck by the idea that science progresses by 'falsifying' earlier theories, meaning that science shuffles off established heuristic models that it has come to find useless, but retains all knowledge that is patently 'true' (more and more the larger part of the whole body of knowledge) because it is borne out by experience. Thomas Kuhn confutes Popper and prefers to see the history of science as a series of 'revolutions' originated by new knowledge that causes scientists to change their way of working, of setting objectives, of evaluating and validating results, and even their language; in other words, it changes science's 'paradigm'.

Each of these approaches has some validity, but it is limited to the specific field to which the approach applies – the history, behavior and psychology of scientists (who, after all, are just people, with brains and minds structured like everyone else's), theories and models in the strict sense – rather than to results established on a precise experimental basis, making no concession to the intellectual superstructures that scientists are (nearly) always tempted to add on (but which are no part of science).

8 Models and artificiality

For philosophers, science is an attempt to construct the world with the human mind – that is, an artificial construct – whereas what science actually does is try to describe the world regardless of the structure of human thought, based simply on the observation of nature.

It is only natural, then, that when philosophers (or anyone else, for the same reason) look at science, they tend to pay more attention to how it is done and to the models that represent it than to science itself. However, models can (if successful) represent only one or a very few aspects of the reality they are intended to reproduce, and sooner or later the facts of nature are bound to falsify the model.

At any rate, it is interesting to note that what is 'artificial' is not only the attempt to reproduce aspects of physical nature, but also the abstract aspects of science and of any phenomenon that can be examined with the human brain and mind. In other words, we can say at this point that construction of the artificial is an 'intellectual intuition' in Kant's sense.

9 Conclusion

General opinion and philosophical thought confuse science and theory/model; indeed, they ignore science and think it consists of models, whether heuristic – these are nearly always necessary to develop new scientific knowledge – or intended to translate scientific truth into terms understandable to the man in the street.

But as science attains higher and higher degrees of 'truth', its models – artificial constructs – become easier and easier to falsify; that is, their artificiality is discovered.

I apologize for the cursory way I've treated the thoughts, analyses and conclusions of scholars, philosophers and scientists here, and described scientific theories and models, but in a short essay I could hardly not do otherwise. My ideas on science are more fully expressed elsewhere (Lanzavecchia and Negrotti), and I shall explain them more thoroughly in a future work (Lanzavecchia 2004).

Bibliography

Accademia Nazionale dei Lincei, *Il Ruolo del Modello nella Scienza e nel Sapere*, in Engl., *The Role of Models in Science and Knowledge*, Accademia Nazionale dei Lincei, Roma 1999.

J.N. Bahcall, R. Davies Jr., P. Parker, A. Smirnov, R. Ulrich (eds.), *Solar Neutrinos. The First Thirty Years,* Westview Press, Boulder 2002.

J. Bell, *Speakable and Unspeakable in Quantum Mechanics*, collected papers on quantum philosophy, Cambridge University Press, Cambridge 1987.

M. Born, *L'expérience et la théorie en physique,* Gauthiers-Villars, Paris 1955; (translation of *Experiment and Theory in Physics*); a short, clear text that gives a convincing idea of what science and theory are

G. Brumfiel, 'Gravity experiment sparks spat between physicists' in *Nature*, 421, 198, 2003.

C. Cellucci 'I modelli, l'analogia e la metafora' in *Accademia Nazionale dei Lincei*, pp 7-25.

A. Cromer, *Uncommom Sense: The Heretical Nature of Science,* Oxford University Press, New York 1993.

G. Galilei, *Dialogo sopra i due massimi sistemi del mondo, tolemaico e copernicano*, Florence 1632, Engl. transl., *Dialogue Concerning the Two Chief World Systems*, Random House Inc., London 2001.

K. Gödel, *On Formally undecidable propositions of Principia Mathematica and Related Systems,* Basic Books, New York 1962, Engl. transl. of the Goedel's article of 1931.

H. Hertz, *Die Prinzipien der Mechanik in neuem Zusammenhang dargestellt*, Leipzig 1894, see also T. Levi-Civita and U. Amaldi, *Lezioni di Meccanica Razionale*, 3 vols., Zanichelli, Bologna 1950-52.

E. Husserl, *Ideas Pertaining to a Pure Phenomenology and to a Phenomenological Philosophy. First Book. General Introduction to a Pure Phenomenology*, Kluwer Academic Publishers, Dordrecht 1982.

G. Lanzavecchia, 'Spazio, tempo e paesaggio', forthcoming in 2003.

G. Lanzavecchia, *Cosa è la scienza?,* forthcoming in 2004.

G. Lanzavecchia, M. Negrotti, *In difesa della scienza. Etica della razionalità e senso comune*, Libri Scheiwiller, Milano 2002.

E. Mach, *The Analysis of Sensations and the Relation of the Physical to the Psychical Source,* Dover Edition, New York 1959. The first complete edition was published in 1906.

E. Mach, *Die Mechanick in hirer Entwickelung historish-kritish dargellstelt*, 1883, Engl. transl., *Mechanics in Its Historical-Critical Development*, 1883.

G.L. Murphy, *The Big Book of Concepts,* MIT Press, Cambridge 2002.

M. Negrotti, *Naturoids. On the Nature of the Artificial,* World Scientific Publishing Co., Singapore 2002.

I. Newton, *Philosophiae Naturalis Principia Mathematica*, 1687.

H. Reichenbach, *The Philosophy of Space and Time*, Dover, 1958.

H. Reichenbach, *Philosophic Foundations of Quantum Mechanics*, Dover, 1944.

H. Reichenbach, *The Rise of Scientific Philosophy,* University of California Press, Berkeley 1951.

L. Russo, *La rivoluzione dimenticata. Il pensiero scientifico greco e la scienza moderna*, Feltrinelli, Milano 1997.

A.S. Sant'Anna, 'An axiomatic framework for classical particle mechanics without force', in *Philosophia Naturalis* 33, 1996, pp. 187-203.

E. Smith, D. Medin, *Categories and Concepts,* Harvard University Press, Harvard 1981.
'The Sun and Solar Wind: A Search for the Beginning', version of the file posted at http://www.genesismission.org/educate/scimodule/S-SWPrOptPDFs/2HowHotIsIt/ModelsInScience-ST-PO.pdf.

Giuseppe Lanzavecchia, LCA – Laboratory for the Culture of the Artificial, Università di Urbino

Roland Müller

Model: The history of the concept and of its use in science

Abstract:

There is much confusion on 'model'. The reason is a lack of 'effort of the concept' and poor knowledge of the colourful history of models. The Latin word 'modulus' was used till 1750. It came three times in the national languages as 'mould', 'module' and 'model'. In the Italian Renaissance models of various kinds were used in art, architecture and anatomy. Since 1555 already mental designs were marked as models and soon model methods were used in science and research. A new wave started 1820 with vivisection and empirical social research, physics and chemistry. The actual model-boom stems from the 1940s and concerns also logic, semantics and mathematics as well as psychology and economics. The development of the computer made possible simulation and modeling, advanced robots and CAD.

Keywords:
model, model muddle, model thinking, model experiments, modelling, mould, module, archetype, icon, representation, copy, pattern, measure, example, mental designs, analogies, metaphors, imagination, holistic thinking, organism, visualization, illustration, classification, simulation, cad (computer aided design), history, science, renaissance, art, architecture, anatomy, physics, chemistry, logic, semantics, mathematics, vivisection, social research, robotics, computer science

1 Is there a 'Model Muddle'?

In 1902 the 'Encyclopaedia Britannica' had a beautiful contribution of Ludwig Boltzmann on 'model', nine years later were added 'artists' model' and 'model-yachting'. In the 1990 edition – 29 volumes! –

there is *no more* article on 'model' neither on 'modelling'. Also in 'Collier's Encyclopedia' (20 volumes; 1993) we do not find 'model' as key word.

In contrast Wolfgang Brezinka (1984) works out 15 different meanings of model on 20 pages, but allows for pedagogy only *one* meaning: the teaching material 'for the illustration or elucidation of an original' (354). In his eyes the other 14 uses are 'useless'.

Despite the programmatic title 'Models' the anthology of 18 essays of the philosopher Max W. Wartofsky from the years 1953-1978 is poor. On only one page he mentions the mechanical models of Maxwell and Thomson, with the criticism of Duhem.

Like many other scholars he does not like the concept 'model' and he therefore speaks 1966 of a 'model muddle' (1979, 1). His solution reads as follows:

> I propose to *collapse the distinction* between models, theories, analogies, and to take all of these, and more besides, as species of the genus representation; and to take representation in the most direct sense of image or copy.

Karl H. Müller (1996, 33) quotes a colleague making fun: 'The modeling paradox can be summarized as an alchemical GIGO-Principle (Garbage in – Gold out)'.

In 1999 Mary S. Morgan and Margaret Morrison report in their introduction to 'Models as Mediators' (1999, 8) after intensive discussions with colleagues for 10 years: 'We have very little sense of what a model is in itself and how it is able to function in an autonomous way'.

However it is of no use to throw in the towel. We just need some 'effort of the concept', already demanded by Hegel, and a fresh look on the colourful history of 'model' and 'models'.

2 Definition and function of models

Each definition of 'model' is insufficient: It covers only a small range of the reach of use. In particular all definitions with the terms 'original', 'reality' and 'image' cannot convince.

Nevertheless one can state as basic meaning: 'A Model is a simplified part of reality or potentiality. It can be material or idealistic, graphic or abstract and describe a has-been, actual or future state'.

A model is – depending on the perspective – either an object of imitation, a copy or replica, draft or substitute, but also an archetype, a pattern or sample form, measure, type or example.

Most models are physical, graphic or intellectual *aids* for the reaching of a goal, what can be: to play or move, control or test, shape and form, plan and decide, register, clarify and impart, explain, as well as to propose and check hypotheses.

A very nice formulation is from Holm Tetens, when he asks in 1986: 'Models in physics. *Ars inveniendi* for strategies of the technical disposal of nature?' Since 1993 one speaks also of 'Models as *mediators*' (Mary S. Morgan, Margaret Morrison 1999, 36, 168-196).

Models often were used playful – in science: *heuristic* (Herman Meyer 1951) or *gnoseologic* (Viktor A. Stoff 1969, 10, 323-328). They serve to create and explore possibilities and to detect by trying out. Therefore Viktor A. Stoff (1969, 32) defines:

> A model is an idealistically visualized or physically realized system. It reflects or reproduces the investigation object and represents it so much, that the study of it provides *new information* about the object.

Models live of the need for visualization. Very often they are attempts to make things vivid (Friedrich Kaulbach 1958; Viktor A. Stoff 1969, 10, 41-45, 279-328; Brigitte Falkenburg 1999), simple and understandable.

Regrettably most books on and of models do not contain any illustrations.

3 The quadruple history of 'modulus'

In all European languages the words modell, model, modèle, modul(e), moule and mould have to be taken together for linguistic and historical reasons. In all these concepts five *fields of meaning* interweave. See Fig. 1.

1. Greek *'metron'* (yardstick, measure, boundary) and Latin 'modus' (respectively in the reduction form: *'modulus'*).
2. Greek *'typos'* (form, sculpture, mould, figuration), Latin *'forma'* (figure, mould, casting) but not: type.
3. Greek *'paradeigma'*, Latin *'exemplar'*, used for small-scale representations of buildings, ships and machines, and in addition, for the (usually male) painter model. The use of architecture models is known since Herodotus. Aristotle uses it in the abstract criticizing Plato's theory of ideas.
4. The weighty Greek words *'idea'* and *'eidos'* (shape, form, idea, Urbild, picture), *'eidolon'* (image, deceit picture) and *'eikon'* (picture). In Latin we have for it *'imago'* and *'effigies'* (picture, ideal, image, conception). Also we have the wide variety of meanings for *'species'* (appearance, picture, idea, classic example, species) and *'simulacrum'* (image, pattern, doll, silhouette, dream vision, fake, characterization).
5. In Plato's dialogue 'Timaios' we find that God creates the human body like a modeller in wax ('keroplastes'). The verbs *'keroplasteo'* and *'proplasso'* were used very rarely for the process of forming. Gr. *'proplasma'* occurs only in the Roman time. For all types of 'wax pictures' the Romans frequently used *'cera'*.

Fig. 1: The Five Fields of Meanings behind 'Model'

3.1 40 BC to 1750: 'modulus' (lat.): measure, figure, architecture model, small scale representation

The Latin word *'modulus'* is the reduction form of 'modus', with the same primary meaning: measure, yardstick. It was not used very much in ancient Rome. Only Vitruvius uses it in his 'Book on Architecture' (approx. 23 BC) various times, mostly as architectural *standard*, namely the half column diameter.

It was the patristic Tertullian (approx. 200 AD), which expanded the meaning on the *plastic figures* of the sculptors, after which for instance a marble sculpture was made.

This *Latin* 'modulus' (like 'exemplar') was used until 1750 in the scholarly world, since 1450 particularly for the architecture model, and in addition, for other small-scale representations of real objects (Leibniz 1903, 163, 223, 596f).

In the national languages we find derivations from 'modulus' since the year 1000 (e.g. Walther von Wartburg 1966; Roland Müller 1983, 1997).

3.2 First development: 'modulus' becomes German 'Model'/'Modul', French 'modle'/'mole'/'moule', English 'mould', Italian 'mòdano'

In German the *loan word* 'Model' – accompanied by 'Modul' – is used since the year 1000 on the one hand in the *abstract sense* as rule, pattern, form, example, on the other hand as *industrial moulds* for things like dragnets and bricks, later documents (e.g. contracts), fabrics and embroideries. Also the forms for printing on textiles and all kinds of hollow forms for casting goods and pastries were called 'Model'.

Quite similarly we have in French: 'modle', 'mole' and 'moule' (otherwise: patron) and in English: 'mould' (otherwise: pattern). Italian 'mòdano' is only in 16th century used as measure, model and mould (otherwise: stampo).

3.3 Second development: 'modulus' becomes Italian 'mòdulo', French and English 'module'

In Italian *'mòdulo'* appears in the 13th century in the sense of Vitruvius' column *measure*. *'Module'* appears 1547 in French and in 1586 in English. We note that in English 'module' is not only used for measure or unit or standard, but also for design, architecture model, image, example.

In the strict sense as measure 'module' held itself through the times, similarly in *physics* as proportionality factor with distortion characteristics (elasticity: Hooke, Young), in *techniques* (with gear wheels) and in *mathematics* (logarithms, congruencies, Abel groups, etc.). For the English language it was Roger Cotes who introduced 1722 the concept of 'modulus'.

In the years 1942-48 the famous Swiss architect Le Corbusier developed the 'Modulor', an architectural standard, based on a man of 6 feet.

The use of 'module' for a standardized construction unit emerges in 1955 in English an expanded rapidly into the other languages.

3.4 Third development: 1366-1417: The complicated birth of the Italian word 'modello'

The Italian word *'modello'* appears in respect to the construction of the Dome of Florence (Rolf Bernzen 1986). Under Francesco Talenti for the first time two wooden models are mentioned. In 1366 the building authority ordered designs for the completion of the cathedral. After short time two drawings and a model from brickwork were available. One of the *drawings* ran under 'desingnum seu modellum'. The three-dimensional structure was named 'Chiesa piccola'.

For technical reasons the construction was not to realize. Fifty years later (1417) Brunelleschi became an advisor, and soon a lot of carpenters were busy producing models for the execution of the dome. These were called now for the first time 'modello', 'modelo', 'modeglio', in Latin 'modello' or 'modellum'. In the year 1420 one was selected. With the help of further models – also for elevators and lifting devices – the construction of the dome was advanced step by step (Howard Saalman 1980). An engineering feat.

3.5 Since 1542: 'modello' arrives in French, German and English

Around 1550 the Italian word 'modello' came:

into French as *'modelle'* or *'modèle'*, into German as *'Modell'* and into English as *'model'* or *'modell'*.

Whereby the variety of meaning expanded rapidly into the present-day abundance.

3.6 Starting from 1555: Mental designs or forms as models

It is remarkable that at that time already *mental* designs were marked as models, such as the *reformation* ('il modello di Martino Lutero' by Giacomo Moronessa 1555), the 'new *Theorick*' of Copernicus (by Thomas Digges 1576), Bacon's *'New Atlantis'* (1627), Descartes' *'Discourse'* (by himself and Huygens) as well as *'The German Principality'* of the Cameralist von Seckendorff (1656).

The philosopher Pascal defined model in this sense as: 'ouvrage d'esprit ou action morale, dont on peut s'inspirer'.

3.7 Rivals of the concept of model

After his birth the concept of the model term immediately got in rivalry to numerous other concepts.

Copernicus used neither in his 'Commentariolus' (approx. 1515) nor in 'De revolutionibus orbium coelestium' (1543) the concepts 'system' and 'model'. But in the title of his main work we find the word *'hypothesis'*. As early as 1540 Joachim Rheticus spoke of the *'systema'* of Copernicus. In 1576 Thomas Digges wrote of the *models* of Ptolome and Copernicus.

Since then further persistent rivals of the concept of model are 'theory' (lat.: theorica), 'philosophy' (philosophia), 'treatise' (tractatus) and 'principles' (principia).

In another perspective important other rival concepts since Antiquity and the Middle Ages were

- *phantasia*, respectively: imago, *imagination*; in German we have also the words 'Einbildungskraft', 'Vorstellung' und 'Anschauung', in English *'idea'*, in French 'idée'.
- *analogia* is richly discussed up to today particularly in catholic theology. Artists and scholars in the Renaissance used analogies virtuous and often. Since that time the use of analogies is common in science and everyday life.
- *metaphors*, mostly in the Scripture, were interpreted by scholars in the Middle Ages.
- *repraesentatio* has, among other uses, been used to declare that the presence of Christ in the eucharisty is not a figment in the imagination (imago in vi-

sione) or in the memory but reality (veritas in repraesentatione), i.e. a real embodiment. St. Thomas used it very often in different senses of sign theory.

4 Reflections on the use of models

Reflections on the use of models start with Xenophanes (540 BC) and Platon. In 1210 AD Robert Grosseteste outlined a model theory in a letter with respect to architecture.

Briefly after 1300 Duns Scotus developed a conception theory and his pupil William of Ockham the conceptualism. Around 1450 Cusanus submitted a picture theory and Leon Battista Alberti a model theory. They were followed by Antonio Averlino, detto Filarete.

Pronounced we find a reflection since the beginning of modern science, for example with Francis Bacon in his doctrine of idols (1620). He distinguishes four kinds of false ideas humans like to form: collective and individual aberrations, public opinion and tradition.

The next wave of consciousness starts with the Cambridge philosopher William Whewell (1840) and the American Scientist Charles Sanders Peirce (1868 to 1903). They were followed since 1875 by the German philosopher Hans Vaihinger, the two Austrian physicists Ernst Mach and Ludwig Boltzmann, the German physicist Heinrich Hertz, the French physicist Henri Poincaré and the Russian physicist Nikolai Alekseevich Umov.

Present reflection on model thinking and the use of models begins in the year 1942. Since 1948 this model-*movement* was so vehement that in 1951 the Viennese physicist Erwin Schrödinger portrayed in his booklet 'Science and Humanism' on five pages the 'nature of our "model"'.

Since 1957 model thinking and the concept of model are discussed at numerous international symposia. In January 1960, August 1994 and October 2000 the IUHPS (International Union for the History and Philosophy of Science) dedicated congresses to the topic 'Model' in Utrecht (Hans Freudenthal 1961), Warsaw (William E. Herfel 1995) and Zurich (Erwin Neuenschwander 2000).

5 The early use of models in science

5.1 Since early 14th century: Anatomical wax models

We can assume continuity from Antiquity to Renaissance workshops using objects of imitation made in clay or from wax, especially to train artistic expression.

The first anatomical models have been shaped in the early 14th century in the course of investigating the vascular system. Molten wax was injected into the vessels, forming a cast that was carefully dissected out from the surrounding tissue. Another of the earliest models was created by none other than Leonardo da Vinci, who injected wax into the ventricular septum of the brain to bring out the detailed structures, and carefully recorded his method in his notebooks.

And more: Leonardo and Michelangelo used not only extensively the word 'modello' but have already made, constructed and used a wide variety of models, not only for buildings of all kinds and for devices but also for artistic purposes, i.e. for drawings and paintings, sculptures and giant sculptures.

5.2 Since 1600: Model experiments

The clergyman Simon Sturtevant described in 1612 in his patent application 'Metallica' the heuristic use of models.

The first model experiments in the sciences made the two Dutchmen Simon Stevin and Cornelius Drebbel, the Englishman William Gilbert (1600 with his terrella, a small iron ball as earth), the Italian Galileo (1638 in the 'Discorsi': experiments on the bend of beam) as well as the German Otto von Guericke (1663 attempts with the Magdeburger hemispheres).

Since 1600 there were constructed ship models for hydraulic experimenting, first in England, then in Holland, France and Russia.

5.3 Since 1600: Hydrodynamic analogies

The ancient Greeks and the artists and scholars of the Renaissance had already used analogies virtuous. In the course of the 'mechanization of the world view' hydrodynamic analogies arose. William Harvey's discovery of the blood circulation (1616) was inspired by hydrodynamic analogies. Also Descartes was stimulated from the hydraulic engineering arts of his time. Giovanni Borelli (since 1666) used likewise analogies from the hydrostatics; he invented the iatromathematics or iatrophysics, a preliminary stage of biophysics.

Whether François Quesnay could be inspired for his model of the economic circulation (1758) of the blood circulation or of hydrodynamics, is disputed. In the physics of electromagnetism since 1820 (Faraday, Gauss, Maxwell) there were used a lot of hydrodynamic analogies.

5.4 Holistic thinking: The organism as analogy

There are also analogies in the other direction. Very popular was the (in fact: idealized) *organism* as a model for social or economic things, sometimes as well as for the universe.

The English statesman and philosopher Thomas Hobbes compared the State (1651) to an 'artificial man'.

Economists and philosophers of the Romantic as Adam Müller (1809) and Franz von Baader (1837) have been inspired by the natural philosophic concept of the organism. The most prominent representative of this kind of thinking was Albert Schäffle. He has tempered the extreme – and therefore wide spread – formulations of his main publication, 'Structure and Life of the Social Body' (1875-78) in its second edition (1896).

5.5 Other analogies

For other analogies the assortment is large: A ball or an apple can be an analogy for the universe or the earth, a clock or a (idealized) machine can be an analogy for the state (Hobbes, Justi), for humans (Lamettrie) or for the universe (Newton-Leibniz), an onion can be an analogy for the soul (Albert Wellek 1950), etc.

6 19th century: Reality, visualization and theory in mathematics and science

6.1 Animals as substitutes for humans

Lamettries 'L'homme machine' (1748) initiated psychological research on animals, and the animal experiments of the universal scholar Albrecht von Haller (1750) laid the foundation for neurobiology.

Since 1820 François Magendie and others regularly made experiments with animals. Pierre Flourens, professor for comparative anatomy in Paris, removed by exstirpation certain brain parts of pigeons and dogs and observed in consequence a decline in performance (1824).

Already in 1831 the neurologist and physiologist Marshall Hall set up guidelines for experiments with animals. The first investigations into isolated hearts took place 1846. And immediately the first protests against vivisection arouse.

6.2 Samples as extracts from society

William Petty was the first to make a sociograph exploration in 1655-56 and soon founded 'political arithmetic'. First statistic work originates from the same time around 1660.

But only about 1740 – when the first consequences of the industrialization began to show up – social sciences took serious the conditions of the working poor. Already around 1795 the Englishmen David Davies and Sir Frederick Morton Eden used a kind of questionnaire to explore them.

A kind of 'social experiments' claiming some representativeness was practiced since 1800 by the so-called utopian socialists Charles

Fourier (1804-1836: 'Phalanstère', and thereof inspired 1844-46: Brook Farm Colony in Massachusetts), Robert Owen (1824-27: 'New Harmony') and Louis Blanc (1840: productive cooperatives).

Empirical social research has been pioneered by the members of the 'Royal Commissions' of 1825, whose efforts culminated in the law for factory inspection of 1833. Their results of research were used still by Marx and Engels.

1895 the Norwegian Anders N. Kiaer presented his idea of a sample based on what he called the 'representative method' for the first time to public. Later he used many representative samples as a basis for statistics, especially with regard to income statistics. In England Sir Arthur L Bowley used 1912-14 the procedure of random sampling.

6.3 Science since 1840: The need for visualization

Another differentiation of the model concept started approximately 1840. On the one hand, it was used for industrial products manufactured in large quantities, on the other hand it played a new role in physics.

In physics there were attempts for visualization. First such concepts were used as picture, illustration, idea, analogy, etc. In 1855 the Scotsman James Clerk Maxwell tried to give a *'geometrical model'* of the physical forces.

Real visualizations of the theory by apparatuses were built since the 1870's in particular by the Englishman Oliver Lodge and the Irishman George Francis Fitzgerald. In the autumn 1892 the German Union of Mathematicians organized an exhibition in Munich of such apparatuses. The extensive 'Catalogue' (ed. Walther Dyck 1892) appeared with a preface of Ludwig Boltzmann.

Already in 1893 the French physicist Pierre Duhem made fun of the efforts of his English colleagues, in particular William Thomson. More than a dozen years later he brought an extended version (70 pages) of his essay as chapter four in his legendary work 'The aim and structure of physical theory' (1906; German: 1908; English: 1914; Italian 1978).

Another Scotsman, Archibald Couper, introduced 1858 to chemistry the graphic visualization by the structural formula with valence lines. The German chemist August Kekulé relied on it and built atomic models and models of molecules from balls and wires 'because of an irresistible need of visualization'. Likewise many mathematicians, e.g. Julius Plücker and Ernst Eduard Kummer, started the plastic modelling of complicated mathematical and geometrical curves and bodies. Gerd Fischer (1986) published two rich and beautiful illustrated volumes.

6.4 The case of William Thomson: How intensive was his search for models?

Generations of historians and theorists of science have quoted the following sentences of Sir William Thomson (later: Lord Kelvin):

> I never satisfy myself until I can make a mechanical model of a thing. If I can make a mechanical model, I can understand It. As long as I cannot make a mechanical model all the way through I cannot understand; and that is why I cannot get the electro-magnetic theory. [...] I can get a model in plain dynamics, I cannot in electro-magnetics.

These sentences are from the beginning of the last of 20 lectures held by Thomson in October 1884 at Johns Hopkins University in Baltimore. A.S. Hathaway made a record in shorthand. Of this in December the same year a 'papyrograph volume' was edited. The careful revision of these texts took Thomson 20 years.

Only 1904 the hardcover edition of the 'Baltimore Lectures on Molecular Dynamics and the Wave Theory of Light' was edited – *without* the sentences quoted above! Why? Thomson had supplemented most lectures, and more, in the years 1901-03 he wrote Nr. 16-20 totally new.

Therefore we can ask if it is fair to quote again and again the shorthand notes and to ignore the final edition of 1904. (The 20 lectures 'in their original form' have been edited by Robert Kargon and Peter Achinstein as late as 1987. Here we find the quoted sentences on page 206).

7 20th century: Models in modern logic, robotics, social sciences, computer science

7.1 Barely used in physics

Surprisingly the model term – apart from the 'atom model' (1913) and model in learning psychology (around 1930) – has been a wall flower nearly half a century after 1900. Most authors did not know what to make with it (e.g. Norman Robert Campbell 1920) or they mention briefly the material models of Maxwell, Thomson and Boltzmann (e.g. Paul Volkmann 1910; Abram Cornelius Benjamin 1937).

Philipp Frank (1927), James Jeans (1933) and Pascual Jordan (1936) thought about *illustrativeness* in physics (also Alvin Diemer 1964; Viktor A. Stoff 1969, 287-298).

For Hermann Weyl (1927) model belongs into the range of logic. Nevertheless the physicist Percy W. Bridgman meant (1927, 53): 'I believe that the model is a useful and indeed inescapable tool of thought, in that it enables us to think about the unfamiliar in term of the familiar'.

Mostly theorists tried to replace model by 'symbol' (e.g. Benjamin) or 'icon', 'isomorphism' and 'analogy' (e.g. Campbell).

In 1963 Mary Brenda Hesse constructed literarily a contrast between the ideas of Pierre Duhem (1906) and Campbell (1920). This is not convincing because no such dispute took place.

7.2 Since 1915: Extension in the modern logic

The model concept was expanded once again by the Viennese philosopher Ludwig Wittgenstein (1921) and the German mathematician Hermann Weyl (1927). In the 1930's logicians as Rudolf Carnap (1934ff), Morris Raphael Cohen and his disciple Ernest Nagel (1934) and Alfred Tarski (1935; 1935/36) saw the model as 'fulfilment' of axiomatic systems and formalized theories.

In the 'Historical Dictionary of Philosophy' we read: 'in logic "model" is a system of areas and concepts, in so far as it fills the axioms of an adequately formulated theory'.

A meanwhile classical definition of Tarski (1953) is: 'A possible realization in which all valid sentences of a theory T are satisfied is called a model of T'.

After World War II model theory expanded further in the logic, semantics and mathematics. One of the first essays in this area, in which we find the expression 'theory of models', is from Alfred Tarski (1954/55). A first introduction to this kind of model theory gave Abraham Robinson in 1963.

7.3 Since 1920: Electrical and electronic robots

Inspired of Karel Čapek's play 'R. U. R.' ('Rossum's Universal Robot', 1920) numerous so-called 'robots' were built – for example from the American engineer Whitman 'Radioman Occultus', a robot soldier. It moves on caterpillar feet, with arms stretched out in front ending in wheels bearing a series of fearsome clubs. 1927 an engineer of Westinghouse, R.J. Wensley, built the domestic robot 'Televox'.

The annual exhibition of the 'Engineering Model Society' in 1928 was opened by a speech of 'Eric', a 'robot in knight's armour' (built by Captain W.H. Richards). Since then each fair and exhibition, particular about its appearance, had its own, often larger-than-life robot (Gary Jennings 1962; Helmut Swoboda 1967, 179ff, 204f; Jasia Reichardt 1978; Brian Morris 1985; Gero von Randow 1997). A whole series 'Sabor I' to 'Sabor V' were built (in 1960) by the Swiss Appenzeller August Huber.

Edwin Garrigue Boring (1946) mentions a series of attempts of renowned psychologists between 1929 and 1938 to design learning robots under titles as: 'A mechanical model of the conditioned reflexes' (H.D. Baernstein, C.L. Hull 1931) and 'A model of the synthesis of conditioned reflexes' (G.K. Bennet, L.B. Ward 1933).

In 1938 the American Thomas Ross built the first scientific robot. It was a small machine, learning like a mouse the way out from a maze by trial and error.

Since then researchers like to tinker around with 'life-imitating' machines. Famous are the electronic turtles 'Elmer' and 'Elsie' of the American-English brain researcher William Grey Walter (1948), the 'Homeostat' of the Englishman W. Ross Ashby (1948) and the 'machina labyrinthea' of the American R.A. Wallace (1952).

Also some theorists of cybernetics and communication as Heinz Zemanek, Norbert Wiener (with his assistant Jerome Wiesner), Claude Shannon and Albert Ducrocq constructed simple cybernetic animals.

The first complete robot system with sensors and optical pathfinder was developed 1968 at the Stanford Research Institute and was called 'Shakey'. It was rebuilt 1971 for other tasks.

The first computer for process control in industry was operative in 1957. Numerical controlled industry robots are in use since 1960.

7.4 1948-1956: Models in psychology and economics

After World War II ingenious models were developed in *learning psychology* (William Kaye Estes; Robert Ray Bush, Frederick Mosteller) as well as in *econometrics* (Tjalling C. Koopmans; Lawrence Robert Klein; Robert W. Solow), *decision theory* (Leonard J. Savage; Ward Edwards; Robert McDowell Thrall, Clyde H. Coombs) and *portfolio theory* (Harry M. Markowitz; William Jack Baumol; James Tobin).

Starting with his dissertation 1950 Patrick Suppes published dozens of contributions concerning models and science in all facets. Anthologies of them appeared in 1979 and 1993.

7.5 Since 1950: Simulation and modelling

In the early 1950's researchers started with simulations on the computer. They tried to introduce mathematical, statistical or stochastic models for the most different processes. In 1958 the AIIE held a first symposium in Baltimore on 'system simulation'. Important impulses came from the RAND Corporation in Santa Monica.

End of the 1970's a bibliography counted over 6000 titles for 'computer simulation' (Per Holst 1979).

Parallel efforts ran to find 'mechanical models' for economic, biological and psychological processes (N.F. Morehouse et al. 1950; O.J. Smith, J.M. Erdley 1951; Arnold Tustin 1953; Donald E. Broadbent 1957).

The concept of *'modelling'* appears somewhat later, on the one hand in the Russian (and the East German translations), on the other hand in English (Eric John Barker 1954; N.L. Irvine, L. Davis 1955; Herbert A. Simon 1961; Kenneth M. Sayre 1963; Richard F. Reiss 1964; George Jiri Klir, Miroslav Valach 1967; Francis F. Martin 1968).

7.6 Since 1960: A revolution – Computer Aided Design (CAD)

Many traditional and beloved techniques of drafting and designing 'models' for buildings, ships or automobiles, electrical circuits or injection molds have been replaced since 1960 by Computer Aided Design (CAD).

This kind of computer use was advanced from US-military research on space travel. Later they made them accessible also for the public. 1964 IBM developed the first CAD Computer, the 'System 2250'. A first general introduction in the field gave 1968 Charles Russell Mischke.

7.7 Not convincing classifications of models

Arturo Rosenblueth and Norbert Wiener (1945) differentiated between *material* and *formal* or intellectual models. Karl Wolfgang Deutsch (1951) and many others followed them; also Viktor A. Stoff (1969, 36-50, 315-323). He sees as material models for instance drafts of houses and cities, layout plans and spatial models of molecules, as thinking models geographic maps and chemical structure formulae.

In 1949 Karl Wolfgang Deutsch introduced the distinction of *mechanistic* (since Newton and Hobbes) and *organismic* (since Rousseau and Burke) models. In 1951 Herman Meyer differentiated scientific models in *mechanistic* ones (e.g. Newton's 'Principia'), *arithme-*

tic (e.g. game of dice) and *axiomatic* ones (e.g. for the kinetic theory of gases).

In their 'Introduction to Operations Research' (1957, 155ff) C. West Churchman, Russell L. Ackoff and E. Leonard Arnoff differentiate between three kind of scientific models: *iconic*, *analogue* and *symbolic*.

Short time later the linguist Max Black (1962, 219-243) differentiated '*scale* models' from '*analogue* models'. The first are mostly reduced images, eventually chemical or social experiments. Black relates them with Charles Sander Peirce's 'icon'.

Analogue models consist of another material than the original; they describe only its structure. They are based on the principle of 'isomorphism'. For Black '*mathematical* models' are nothing else than mathematical 'treatments' without explaining power.

For '*theoretical* models' Black refers to Maxwell and Thomson and states, they are only verbal descriptions, 'half understood metaphors'. They lead to mystifications and confusion of concepts. Black considers the use of such models as a detour. Finally he introduces the term *'archetype'* for 'root metaphor', 'basic analogy' or 'ultimate frames of reference'.

Gerhard Frey introduced 1960 the distinction of *symbolic* (e.g. a set of equations), *primarily iconic* (e.g. a wiring diagram) and *secondarily iconic* models (e.g. the theories of the classical mechanics). A symbolic model, to which there is a completely copying iconic model, is an iconic-symbolic model. In contrast the Heisenberg matrix mechanics and the Schroedinger wave mechanics are not-iconic-symbolic models.

The most elaborate classification is the German mathematicians Herbert Stachowiak comprehensive general theory of models ('Allgemeine Modelltheorie', 1973). It was widely ignored.

Bibliography

A.C. Benjamin, *The logical structure of science*, Paul, Trench, Trubner, London 1936.

L. Boltzmann, 'Model', in *Encyclopaedia Britannica*, 'The Times' Printing House 10th Edition, volume XXX, London 1902, pp. 788-791, word-for-word reprinted in the 11th Edition, Cambridge University Press, Cambridge 1911, vol. 18, pp. 638-640, further in *Theoretical Physics and Philosophical Problems*, Reidel, Dordrecht 1974, pp. 214ff.

R. Bernzen, *Die praktische und theoretische Konstruktion des Modellverfahrens. Ein Beitrag zur Frühgeschichte der neuzeitlichen Wissenschaft*, Peter Lang, Frankfurt am Main 1986.

Max Black, *Models and Metaphors. Studies in Language and Philosophy*, Cornell University Press, Ithaca (NY) 1962, 3. ed. 1966.

E.G. Boring, 'Mind and Mechanism', in *American Journal of Psychology* 59, 1946, pp. 173-192.

W. Brezinka, '"Modelle" in den Erziehungstheorien', in *Zeitschrift für Pädagogik*, 30. Jg., 1984, pp. 834-858.

P.W. Bridgman, *The Logic of Modern Physics*, Macmillan, New York 1927, again 1960, reprint Ayer, Salem (NH) 1993, Ger. transl., *Die Logik der heutigen Physik*, Hueber, München 1932.

D.E. Broadbent, 'A Mechanical Model for Human Attention and Immediate Memory', in *Psychological Review* 64, 1957, pp. 205-215.

N.R. Campbell, *Physics. The Elements*, The University Press, Cambridge 1920, unchanged reprint *Foundations of Science. The Philosophy of Theory and Experiment*, Dover, New York 1957.

R. Carnap, *Logische Syntax der Sprache*, Julius Springer, Wien 1934, 2. ed. 1968, Engl. transl., *The Logical Syntax of Language*, Paul, Trench, Trubner, London 1937, 2. ed. 1949, reprint Routledge, London 2000.

R. Carnap, *Philosophy and Logical Syntax*, Paul, Trench, Trubner, London 1935, reprint AMS Press, New York 1979.

C.W. Churchman, R.L. Ackoff, E.L. Arnoff, *Introduction to Operations Research*, Wiley, New York 1957, 13. ed. 1969.

K.W. Deutsch, 'Some Notes on the Role of Models in the Natural and Social Sciences', in *Synthese* 7, 1948-1949, pp. 506-533.

K.W. Deutsch, 'Mechanism, Organism, and Society. Some Models in Natural and Social Science', in *Philosophy of Science* 18, 1951, pp. 230-252 (quoted 230) (reprint 1967).

A. Diemer, 'Anschaulichkeit', in *Grundriss der Philosophie*, 2. Bd., Hain, Meisenheim 1964, pp. 574ff.

P. Duhem, *La théorie physique. Son objet et sa structure*, Chevalier & Rivière, Paris 1906, 2. enlarged ed. 1914, numerous reprints to Paris: Vrin 1997, further reprint Minerva, Frankfurt am Main 1985, Ger. transl., *Ziel und Struktur der physikalischen Theorien*, Barth, Leipzig 1908, reprint Meiner, Hamburg 1978, again 1998, Engl. transl., *The Aim and Structure of Physical Theory*, Princeton University Press, Princeton (NJ) 1954, reprints Athenaeum, New York 1962 and 1981, Princeton 1991, It. transl., *La teoria fisica. Il suo oggetto e la sua struttura*, Il Mulino, Bologna 1978.

W. Dyck (ed.), *Katalog mathematischer und mathematisch-physikalischer Modelle, Apparate und Instrumente*, C. Wolf u. Sohn, München 1892, reprint Olms, Hildesheim 1994.

B. Falkenburg, 'Sprache und Anschauung in der modernen Physik', in H.E. Wiegand (ed.), *Sprache und Sprachen in den Wissenschaften*, de Gruyter, Berlin 1999, pp. 89-118.

G. Fischer (ed.), *Mathematische Modelle. Aus den Sammlungen von Universitäten und Museen*, 2 vol., Vieweg, Braunschweig 1986.

P. Frank, 'Über die "Anschaulichkeit" physikalischer Theorien', in *Naturwissenschaften*, Bd. 16, 1928, Heft 8, pp. 122ff.

H. Freudenthal (ed.), *The Concept and the Role of the Model in Mathematics and Natural and Social Sciences*, Proceedings of the Colloquium Sponsored by the Division of Philosophy of Sciences of the International Union of History and Philosophy of Sciences, Organized at Utrecht, January 1960, Reidel, Dordrecht 1961.

G. Frey, 'Symbolische und ikonische Modelle', in H. Freudenthal (ed.) 1961, pp. 89-97.

W.E. Herfel et al., *Theories and Models in Scientific Processes*, Proceedings of AFOS '94 Workshop, August 15-26, Madralin, and IUHPS '94 Conference, August 27-29, Warszawa, Rodopi, Amsterdam 1995.

M.B. Hesse, *Models and Analogies in Science*, Sheed and Ward, New York, London 1963, 2. ed. with two new chapters, University of Notre Dame Press, Notre Dame (IN) 1966, second printing 1970.

P. Holst, *Computer simulation 1951-1976. An index to the literature*, Mansell, London 1979.

N.L. Irvine, L. Davis, 'Simulation by Modeling', in *Proceedings (AFIPS) Western Joint Computer Conference*, vol. 7, 1955, pp. 13-15.

J.H. Jeans, *The new background of science*, Cambridge University Press, London 1933, 2. ed. 1934, reprint 1959, Ger. transl., *Die neuen Grundlagen der Naturerkenntnis*, Deutsche Verlagsanstalt, Stuttgart 1933, 2. ed. 1934.

G. Jennings, *March of the Robots. From the manikins of antiquity to the space robots of tomorrow*, Dial Press, New York 1962.

P. Jordan, *Anschauliche Quantentheorie. Eine Einführung in die moderne Auffassung der Quantenerscheinungen*, Springer, Berlin 1936.

R. Kargon, P. Achinstein (eds.), *Kelvin's Baltimore Lectures and modern theoretical Physics. Historical and Philosophical Perspectives*, Massachusetts Institute of Technology, Cambridge (Mass.) 1987.

F. Kaulbach, 'Die Anschauung in der klassischen und modernen Physik', in *Philosophia naturalis*, 5, 1958, pp. 66-95.

G.J. Klir, M. Valach, *Cybernetic Modelling,* Illiffe, London, S.N.T.L., Prague 1967.

G.W. Leibniz, *Opuscules et fragments inédits*, ed. L. Couturat, Olms, Hildesheim 1966 (reprint of the edition Paris 1903).

F.F. Martin, *Computer Modeling and Simulation*, Wiley, New York 1968.

H. Meyer, 'On the heuristic value of scientific models', in *Philosophy of Science*, 18.2, 1951, pp. 111-123 (reprint 1967).

C.R. Mischke, *An introduction to computer-aided design. Fundamentals of engineering design*, Prentice-Hall, Englewood Cliffs 1968.

M.S. Morgan, M. Morrison (eds.), *Models as Mediators. Perspectives on Natural and Social Science*, Cambridge University Press, Cambridge 1999.

B. Morris, *The World of Robots*, Gallery Books, New York 1985, Ger. transl., *Die Welt der Roboter*, Umschau-Verlag, Frankfurt am Main 1986.

K.H. Müller, 'Epistemic Cultures in the social sciences – The Modelling Dilemma Dissolved', in R. Hegselmann, U. Mueller, K.G. Troitzsch (eds.), *Modelling and simulation in the social sci-*

ences from the philosophy of science point of view, Kluwer Academic Publications, Dordrecht 1996, pp. 29-63.

R. Müller, 'Zur Geschichte des Modelldenkens und des Modellbegriffs', in H. Stachowiak (ed.), *Modelle – Konstruktion der Wirklichkeit*, Fink, München 1983, pp. 17-86.

R. Müller, 'The Concept of Model and Its Triple History', in E. Neuenschwander (ed.) 2000, pp. 7-19.

E. Neuenschwander (ed.), *Scientific Models: Their Historical and Philosophical Relevance. Book of Abstracts*, XIIIth DHS-DLMPS Joint Conference, 19-22 October 2000, Geschichte der Naturwissenschaften, Universität Zürich 2000.

G. von Randow, *Roboter. Unsere nächsten Verwandten*, Rowohlt, Reinbek 1997, paperback 1998.

J. Reichardt, *Robots: Fact, Fiction and Prediction*, Thames & Hudson, London 1978.

R.F. Reiss (ed.), *Neural Theory and Modelling*, Proceedings oft the 1962 OJAI Symposium, Stanford University Press, Stanford (CA) 1964.

A. Robinson, *Introduction to Model theory and to the Metamathematics of Algebra*, North-Holland, Amsterdam 1963, 2. ed. 1965.

A. Rosenblueth, N. Wiener, 'The Role of Models in Science', in *Philosophy of Science*, 12, 1945, pp. 316-321.

H. Saalman, *Filippo Brunelleschi. The Cupola of Santa Maria del Fiore*, Zwimmer, London 1980.

E. Schrödinger, *Science and Humanism*, Cambridge University Press, Cambridge 1951, Ger. transl., *Naturwissenschaft und Humanismus*, Deutike, Wien 1951, see pp. 32-36.

H.A. Simon, 'Modeling human mental processes', in *Proceedings (AFIPS) Western Joint Computer Conference*, vol. 19, 1961, pp. 111-120.

V.A. Stoff, *Modellierung und Philosophie*, Akademie-Verlag, Berlin 1969 (from Russ. Nauka, Moskau 1966).

P. Suppes, 'Models of Data', in E. Nagel, P. Suppes, A. Tarski (eds.), *Logic, Methodology and Philosophy of Science*, Proceedings of the 1960 International Congress, Stanford University Press, Stanford (CA) 1962, pp. 252-261.

P. Suppes, *Studies in the methodology and foundations of science. Selected papers from 1951 to 1969*, Reidel, Dordrecht 1969.

P. Suppes, *Models and Methods in the Philosophy of Science. Selected Essays*, Synthese Library, vol. 226, Kluwer, Dordrecht 1993.

H. Swoboda, *Der künstliche Mensch*, Heimeran, München 1967.

A. Tarski, 'Der Wahrheitsbegriff in den formalisierten Sprachen', in *Studia Philosophica*, 1, 1935, pp. 261-405 (original thesis in Polish, Warsaw 1933), Engl. transl. in *Logic, Semantics, Metamathematics*, Clarendon Press, Oxford 1956, pp.152-278, German reprint in K. Berka, L. Kreiser (eds.), *Logik-Texte*, Akademie-Verlag, Berlin 1971.

A. Tarski, 'Grundzüge des Systemenkalküls I/II', in *Fundamenta mathematicae*, 25/26 (1935/36), pp. 503-526, pp. 283-301, Engl. transl. in *Logic, Semantics, Meta-mathematics*, Clarendon Press, Oxford 1956, pp. 343-383.

A. Tarski, 'Contributions to the theory of models', in *Indagationes Mathematicae*, 16, 1954, pp. 572-588, 17, 1955, pp. 56-64.

H. Tetens, 'Modelle in der Physik. Ars inveniendi für Strategien der technischen Verfügung über die Natur?' in C. Burrichter, R. Inhetveen, R. Kötter (eds.), *Technische Rationalität und rationale Heuristik*, Schöningh, Paderborn 1986, pp.169-184.

W. Thomson (Lord Kelvin), *Baltimore Lectures on Molecular Dynamics and the Wave Theory of Light. Followed by twelve Appendices on allied Subjects*, C.J. Clay and Sons, Cambridge University Press Warehouse, London 1904, Ger. transl., *Vorlesungen über Molekulardynamik und die Theorie des Lichts*, Teubner, Leipzig 1909.

A. Tustin, *The mechanism of economic systems. An approach to the problem of economic stabilisation from the point of view of control-system engineering*, Heinemann, London 1953, 2. ed. 1957.

P. Volkmann, *Erkenntnistheoretische Grundzüge der Naturwissenschaften und ihre Beziehungen zum Geistesleben der Gegenwart*, Teubner, Leipzig 1896, 2. enlarged ed. 1910.

W. von Wartburg, *Französisches Etymologisches Wörterbuch*, Zbinden, Basel 1922-1967.

M.W. Wartofsky, *Models. Representation and the Scientific Understanding*, Reidel, Dordrecht 1979.

A. Wellek, *Die Polarität im Aufbau des Charakters. System der Charakterkunde*, Francke, Berlin 1950, 3. ed. 1966.

H. Weyl, 'Philosophie der Mathematik und Naturwissenschaften' in *Handbuch der Philosophie*, Bd. 2, Oldenbourg, München, Berlin 1927, pp. A1-162.

L. Wittgenstein, 'Logisch-philosophische Abhandlung', in *Annalen der Naturphilosophie*, 14, 1921, *Tractatus logico-philosophicus*, German-English, with an introduction by Bertrand Russell, Paul, Trench, Trubner, London 1922, only in German, Suhrkamp, Frankfurt am Main 1960, 26. ed. 1999, *Logisch-philosophische Abhandlung. Kritische Ausgabe*, Suhrkamp, Frankfurt am Main 1998.

Roland Müller, lecturer at the Chair of Methodology Psychological Institute, University of Zurich

Giuseppe Padovani

An Implicit Model in Sociology: Raymond Boudon

Abstract:

The aim of this essay is to show how the model of action *proposed by Raymond Boudon develops an ambiguous concept of rationality, which, in his latest works, comes to assume a cognitive connotation somewhat akin to J. Piaget's psychology, abandoning the 'lesson' of the classic sociologists of* social action*, and in particular that of M. Weber. Boudon's sociology leads to an outcome that is in some ways paradoxical, leaving us with a question: how do we explain why the ideas, beliefs and feelings of the actor, whilst resulting from correct forms of logical reasoning, are nevertheless fragile, erroneous, not objectively valid?*
The model proposed by Boudon, inspired by G. Simmel, would have the intention of explaining how this is possible; but in this way, the focus is moved from the analysis of the reasons for, and the purposes of, the social action of the individual in given circumstances, to the logic of the underlying reasoning, which, as such, belongs to the universality of cognitive rationality.

Keywords:
model, social interaction, cognitivism

1 Introduction

The classics, as it were, of sociology – A. Comte, E. Durkheim, M. Weber, G. Simmel, and V. Pareto – upon whose thought R. Boudon's discourse rests – have indicated that it is not possible to distinguish the *object* of the social sciences from the *theory* that constitutes it. Sociology moves from a *pre*-comprehension, which guides research towards social reality and opens it to intelligibility. Thus Weber clearly delimits the conceptual framework within which sociological discourse, be it theoretical or methodological, is explained. As is well known, he sets social action as the object, and individualism as the methodological principle, of sociological research.

I retain that it is not possible to speak of models in sociology, whilst setting aside the frame of reference. In sociology, a model is just such to the extent to which it satisfies two basic requirements: it must be adequate to the doctrine, and, at the same time, adequate to the method. Even when the sociologist must make use of a quantitative-cum-statistical or a logical-cum-abstract model – as is often the case when collecting, processing and interpreting data – that model must comply with, and correspond to, the requirements of sociological explanation, i.e. of the theory that guides the cognitive enterprise, and which is thus put to the test in the research.

In his or her work, the sociologist may formulate *ad hoc* models, or use models borrowed from other disciplines (on the theory of models, see also the work of Bruschi, 1971, and the collection of essays by Boudon himself, 1971). Thus, in sociology, we often find models taken from economy, linguistics, cybernetics, game theory, and so on. These models, as exemplified by Boudon, may be classified according to 'leur finalité, c'est-à-dire du point de vue de l'objectif recherché par le sociologue qui les utilise. Ils peuvent en outre être classés en fonction de leurs caractéristiques logiques' (Boudon 1982, p. 393). In other words, in sociology, a model may be used for different purposes, but it must always satisfy the criteria dictated by sociological theory and its epistemological premises. This is particularly true when we consider – as does Boudon himself – the lessons of Weber and Simmel, according to which it is impossible to explain an act or a phenomenon realistically, as this is a *reconstruction* arising from a model. In this sense, it is useful to consider the definition of model given by Boudon himself:

> Un modèle c'est une représentation idéalisée dont on présume qu'elle peut permettre de mieux comprendre certaines situations réelles, à condition de prendre conscience des simplifications que le modèle introduit. Un modèle a la double propriété d'être *général* – dans la mesure où il peut s'appliquer à des contextes spatio-temporels divers – et *idéal* dans la mesure où il ne s'applique textuellement à aucune réalité concrète (Boudon 1982, p. 523).

A model is neither a law nor a theory. Laws and theories make use of models in order to analyse phenomena. A law is an empirical statement with universal value, while a model is an ideal statement with general value. A model is open, so to speak, to the real, because it selects and is orientated towards those phenomena that must be observed, and predisposes them to explanation. Moreover, the model satisfies a heuristic requirement, and it is from this point of view that we shall speak of the models, and specifically the implicit models, of Boudon.

2

In conceptually delimiting the frame of reference of sociological models, Weber, as we know, sets social action as the object of sociological knowledge. However, Boudon develops only certain aspects of this frame of reference, and others he does not consider at all. The radicalisation of the methodological principle of individualism urges him to focus his attention on the elementary individual actions from which social phenomena originate. Although the action of the individual develops 'à l'intérieur d'un système de *contraintes*', of which the actor has some knowledge, it is not, as holism maintains, the effect of the conditioning produced by social structures.

If, on the one hand, action is not reducible to the effects of conditioning, then, on the other, the actor's preferences and the means at his disposal, or which he believes to be at his disposal, are connected to social structures. In Boudon's judgement, it is exactly this aspect that renders the action comprehensible. For the observer, understanding the action means performing some cognitive work which leads him to place himself in the actor's shoes, so to speak:

> La relation de compréhension qui *peut* s'instaurer entre l'observateur et l'acteur n'est pas immédiatement donnée. Elle suppose généralement de la part de l'observateur un travail d'observation et un souci de distanciation pour *comprendre* l'action d'autrui, l'observateur doit prendre conscience des différences qui distinguent sa situation propre de celle de l'observé (Boudon 1982, p. 5).

This implies that whatever cultural distance separates the observer from the observed, the former will, in principle, be able to understand the action of the latter. Therefore, individual action implies that there be some invariant elements *('éléments invariants')* with respect to change and diversity of the cultural context. These invariant elements – i.e., man's nature *('la nature humaine')*, common to the observer and to the observed – make comprehension possible in spite of differences between cultures.

Upon this basis Boudon anchors the assumption of the rationality of the action and the methodological principle of investigating the reasons – the good reasons – that orientate the actions of individuals, whatever (and indeed, in Boudon's latest works, notwithstanding) their socio-cultural background. In Boudon's model, the reasons for actions, although related to social constrictions, are, in the final analysis, disengaged from them. Thus, as we shall later see, the actions of the individual can aggregate and combine at the macro-social level, thanks to the universality of reason that underlies them.

The actor always has 'good subjective reasons', in the sense that he has personal motives, for doing what he does, or for believing what he believes, or he always has 'good reasons', in the sense of arguments – reasoning – for justifying what he does or what he believes. Boudon does not always distinguish clearly between these two levels. Moreover, in his latest works he tries ever more insistently to characterise the rationality of the actor's choices without taking into account the social background. Thus, the difference between Boudon's concept of social action on the one hand, and that of Weber on the other, becomes increasingly marked. In his latest works, Boudon speaks only of action, or human action. This reduction allows him to put forward three models of action: sociological, comprehensive, and cognitive. Thus, if we wish to explain a value-judgement according to each of these models:

> [...] [le modèle] *sociologiste* veut que l'acteur social ait tendance à endosser les jugements des valeurs en vigueur dans le milieu social qui est le sien. L'acteur croit, nous dit Durkheim, à ce qu'on croit autour de lui parce qu'il s'identifie au groupe et que ses valeurs s'imposent à lui. [...] Le modèle *compréhensif* veut que l'acteur social endosse les jugements de valeur qui font sens pour lui. Expliquer une croyance, normative ou positive, c'est en saisir le

> *sens* pour l'acteur. Lorsqu'on précise que 'comprendre' une croyance, en retrouver le 'sens', c'est reconstruire les *raisons* que l'acteur a de l'endosser, l'on peut parler de modèle *cognitiviste* (Boudon 1994, p. 22).

Weber instead speaks of social action. An action is social when the subject orientates the subjective sense of his own action towards that of the action of another. The action is comprehensible because it carries a sense that, although subjective, is intelligible to the other and to the observer. The rationality of an action is made possible by its intelligibility, by the fact that it is sensed as a human action. For Weber, the notion of sense is wider than that of rationality. In general, for him, the term 'sense' indicates that the actor, as a man, performs actions that have a sense, a meaning, because they are performed with a view to realising something he would like realised, or in relation to something else that has sense for him. Boudon prefers instead to speak of 'reason', as he considers the Weberian term 'sense' to be ambiguous. However, because he replaces the term 'sense' with the term 'reason', he needs to develop a new concept of the rationality of action.

Finally, for Weber, the concept of social action is coextensive with that of social relationship. (We recall that an action is social insofar as it is orientated according to the sense of the action of the other.)

3

Boudon has developed a convincing theory of action, or *actionist* sociology *('actionniste')*, as a paradigm for understanding and explaining social phenomena. According to this paradigm, every social phenomenon is the aggregated result of the actions, attitudes, beliefs and behaviour of individuals. (A paradigm, as Boudon reminds us himself, is made up of the same principles that orientate social research.) The explanation of a social phenomenon consists in tracing the meaning, or the 'good reasons', that individuals attribute to their own actions, which have contributed, more often than not independently of one another, to the composition of that particular social phenomenon.

In other words, the sociologist, in order to explain a social phenomenon, must conform to the principles of methodological individualism and of the rationality of the actor; he must understand the reasons that social actors have for doing what they do and believing what they believe. Indeed, the social actor always has 'good reasons' for doing what he does and believing what he believes (as he is the deviser of the action or belief). In principle, these reasons should be comprehensible to the sociologist, although not always easily. In the final analysis, the sociologist must understand, or discover, the reasons why the social actor does what he does.

However, one cannot but observe that these reasons are, for the most part, reconstructed by the sociologist. Indeed, such 'reasons' are not sought by questioning the actors, or by listening to them, but are instead inferred. It would be necessary, as suggested by G.G. Granger, to distinguish between *'rationalité épistémique'* and *'rationalité pratique'* – that is, between the rationality of the observer and that of the actor. These two types of rationality are not superimposable.

Boudon, rather, develops a model of rationality that renders intelligible the behaviour of social actors – a model that is placed, so to speak, on the side of the interpreter, i.e. of the sociologist. When he speaks of 'reasons', 'beliefs', 'value-judgements', 'moral sentiments', and 'sentiments of justice', he refers in general to the cognitive process of justification, to the argumentation, which, according to the interpreter, rules or supports these reasons, beliefs, etc. However, Boudon is not interested in contents – in the use that actors make of their beliefs, feelings, and so on – but rather in the cognitive process, which is, in a certain sense, independent of them. Indeed, as he explicitly maintains:

> Le modèle cognitiviste propose plus simplement d'admettre que le caractère contraignant des jugements de valeur et des sentiments moraux provient de ce qu'ils se fondent sur des raisons qu'on peut qualifier 'transubjectives' (Boudon 1994, p. 24).

The cognitive model is *'transculturelle'* and *'transcontextuelle'* in the same way that scientific reasoning is:

> La pensée ordinaire comme la pensée scientifique abordent la complexité du réel à l'aide de conjectures qui, parce qu'elles sont fondées sur des bonnes raisons ont une valeur trans-subjective et ont vocation à être partagées (Boudon, 1997, note 2, p. 29).

However, cognitive rationality – that is, the ability to deduce reasons from beliefs, values and feelings – is never directly observed, but is *reconstructed.*

In Boudon's work we find, as we try to show here, an implicit distinction between the content of reasoning and the formal aspect of reasoning. Boudon makes use of this distinction to hypothesise that if individuals are found to have the same ideas and beliefs, this is due to the fact that, independently of one another, they follow the same process of argumentation. If this were so, however, then many social phenomena of aggregation would have a cognitive explanation rather than a social one.

From these assumptions it follows that:

1 The sociologist must abandon the search for the causes underlying the behaviour and beliefs of the actor, whenever such a search reveals causes that differ from the actor's own rational causes (postulate of the social actor's rationality).
2 The sociologist must go beyond the traditional classification of social actions, in economy and sociology – as rational and irrational, logical and illogical, objective and subjective – to consider social actions as being moved by reasons, and not determined causally, in some way that escapes the actor's consciousness.
3 To this end, the sociologist must formulate a concept of rationality that is consistent with the social action of free social actors. In actionist sociology, an action is judged rational whenever it is possible to understand the 'reasons' of the actor; which is tantamount, in general, to saying 'the fact that actor *X* has behaviour *Y* is *comprehensible*: in effect, in his situation, he has *good reasons* for doing *Y*' (Trattato, p. 42).

Boudon retains that a semantic definition of rationality will suffice – a definition that goes beyond both the *broad* conception of rationality put forward by Karl Popper (cf. Popper 1967) – according to which it is possible to consider an action rational whenever the actor can provide reasons for it, although it is not important which ones – and the *narrow* conception of rationality, according to which one's behaviour is rational only when it is based on objectively founded reasons.

The advantage of a semantic definition of rationality, argues Boudon, is that it does not restrict the actor's rationality to only a formal and objective dimension (the narrow definition), and it also escapes the excesses of the broad definition, thus allowing us to distinguish between different types of rationality. In short, according to a semantic definition, we regard an action *Y* as rational when '*X* had good reasons for doing *Y*, because [...]', where 'because' corresponds to some interest, or to the means of achieving a goal, or to a normative principle, a habit, a belief, and so on. Moreover, this definition has the advantage of extending the concept of rationality to beliefs:

> 'X avait des bonnes raisons de faire Y, car il croyait que Z [...]' a un sens si l'on peut la compléter par une suite de propositions faisant-elles aussi sens, telles que: 'car Z était vrai', 'car Z était vraisemblable', 'car Z était plausible', plus généralement 'car X avait des bonnes raisons de croire Z, car [...]', etc. (Boudon 1990, p. 405).

Once rationality has been defined precisely, it comes to develop a methodological function similar to that of individualism, according to which it is possible to interpret any type of social action. Indeed, in contrast to economics, for example, sociology presents several different types of rationality. Thus, Boudon goes on to show that ideologies, beliefs, sentiments, value-judgements and norms are reducible to cognitive rationality. However, if this assumption is to be plausible, the definition of rationality must embrace all types of action, from the rational actions of the scientist to the actions of each individual engaged in everyday life.

4

In order to explain *how* and *why* a given actor adheres to a certain idea or belief, Boudon proposes a model of analysis of social actions. His model, on the one hand, is a critique and a revision of the formulations of cognitive sociology, whilst, on the other, it represents an abandonment of the classical concepts of social relations as proposed by M. Weber and G. Simmel.

Concerning the first aspect, Boudon criticises the distinction between logical and illogical actions. Usually, he argues, we turn to an explanation of action by means of reasons (type I), or by means of non-rational causes (type II), which may be passions (type IIa), psychical factors (type IIb), or sentiments (Type IIc). In type II explanations, 'les croyances sont interprétées comme l'effet non de raisons, mais de causes non accessibles à l'observation' (Boudon 1990, p. 29). Because the acting subject's reasons are beyond the reach of the observer, the observer is urged to consider them illusory or irrational, and therefore as an effect of some causes:

> Dès qu'une croyance paraît étrange, bizarre, fausse, discréditée, l'explication qu'on tend spontanément à donner du fait que certains y adhèrent relève souvent du type II, l'on recherche les causes plutôt que les raisons et l'on tend à considérer les raisons que le sujet s'en donne comme de simples illusions dépourvues d'effet réel sur ses croyances (Boudon 1990, pp. 29-30).

Instead, we must find a theory capable of showing how an explanation of type II may be replaced, advantageously, by an explanation of type I – this being widely adherent to the principles of methodological individualism and the rationality of the actor. This does not mean that individual reasons lack objective ground. We are interested in showing that the good reasons an individual has for believing something, however subjective such reasons might be, are not 'arbitraires, ni liées à la singularité du sujet. La question est *évidentement* de savoir comment des raisons peuvent être à la fois *bonnes* et objectivement non fondées' (Boudon 1990, p. 45), where 'arbitrary' means that the reasons are supported by a reasoning, and 'subjective' means individual, but not accessible to the observer's comprehension (principle of rationality).

Now, explanations of type II, i.e. not based upon the rationality principle, but on the other socio-centred (or comprehensive) models, are easily transcribed according to an explanation of type I – as Boudon believes himself to have shown. How does it come, then – asks Boudon – that explanations of type II are usually regarded as '*more profound*, more capable of analysing the real in its complexity, than are explanations of type I?' Is it not perhaps *curious* when these explanations, more often than not, are 'both superficial and false?'.

These questions would seem strange coming from a sociologist who aims at explaining social phenomena. Do they not amount to abandoning the classic formulations of Weber and Simmel that inspired Boudon's model? If an individual who acts always has good reasons, of cognitive order, for believing what he believes, then it is useless to look for reasons other than those he has for doing what he does or believing what he believes. However, this assumption does not take into account, nor does it say anything about, the fact that one's individual beliefs are not always a product of one's reasoning, even if one has reasons for adhering to them, accepting them, and justifying them. Above all, though, it does not take into account the fact that one's beliefs have to be translated into action, and to take on, as it were, the materiality of the social, of the relationship with the other, or the others (compare J.R. Searle's 'factives' – Searle 2001).

Let us consider the example of the wizard proposed by Boudon in several parts of his work. It is necessary to note that (1) although the wizard's beliefs lack objective basis, and notwithstanding the fact that the wizard has good reasons for believing what he believes, it is possible that his beliefs are shared with others; and (2) just as importantly, he is a *wizard*, and he believes and applies his beliefs as a wizard, a role acknowledged by others, otherwise he would not be a wizard.

In sociology, an explanation cannot limit itself to recognising that individuals have good reasons for believing what they believe in compliance with only cognitive operations. It must also explain how these reasons are formed and maintained, and how, for individuals, and not only for single individuals, they are good reasons for holding the beliefs. Now, Boudon, when he speaks of rationality, does not refer to this rationality, but to the rationality of the argumentation used by the actor for the purpose of logically justifying his beliefs. Indeed, the wizard, Boudon maintains, is able to justify his beliefs in the same way as the scientist justifies his scientific beliefs. If, instead, one does not refer to logical reasoning, but to the content – to the position or conception of the world, or of the experience implicit in the belief –

then the frame of reference consequently changes, as in such a case the wizard's beliefs have a manifestation at the social level, being as one with his actions, and not just at the cognitive level.

Boudon is convinced that the diffusion of reason-based explanations of behaviour is due also to the change in the theory of knowledge that took place with Kantian criticism. Explanations of type II are based on the classical theory of knowledge, whereby knowledge is a *reflection* of reality (knowledge of): a belief is false because it is the effect of some perturbing causes. With Kant's revolution, knowledge becomes 'knowledge of the fact that' (knowledge that), according to the language of R. Rorty (cf. Rorty 1980) conditioned by the *a priori* categories of the knowing subject in a way that 'introduces a symmetry between adhering to a true idea and adhering to a false idea'.

Both Weber and Simmel, as we know, are inspired by Kantian criticism. Simmel in particular, according to Boudon, develops a theory that is capable of explaining the illusion of beliefs, but which does not deny the subject the use of normal procedures of thought. The development of a theory capable of explaining a phenomenon takes place through the formulation of explicit propositions that articulate the reasoning with other implicit propositions that remain unnoticed:

> Or, suggère-t-il [Simmel], il peut fort bien se produire que la structure de la théorie soit modifiée lorsqu'on explicite ces propositions implicites. Il peut aussi arriver que les conclusions que nous tirons de la théorie soient bouleversées dès lors que les propositions implicites sont mises sur la table (Boudon 1990, p. 59).

5

The thesis that Boudon must demonstrate is that a belief, even if weak or false, is nevertheless the product of an argument on the part of the social actor (who has his 'good reasons'). In support of this thesis, Boudon develops a model that he attributes to G. Simmel. His model would make it possible to interpret social actions in terms of rationality, however limited or cognitive, and thus to overcome the classical distinction between objective and subjective rationality. Above all, however, as we have hinted above, the model allows us to reject any type of causal explanation of actions.

The salient points of the model proposed by Boudon are the following:

1 All knowledge, as well as all judgement, rests on and is made possible by the use of the *a priori*.
2 *A priori* categories (not identified with Kant's categories) perform the same function, and can also constitute a source of errors.
3 A justificational argument, making use of the *a priori*, may encounter the discordance between the way in which the subject perceives his own reasoning, and reasoning as such.
4 *A priori* categories, which act as hypotheses or conjectures of some process of thought, are both implicit and generally unconscious, i.e. *meta-conscious*.

Any process of thought, then, whether it be cognitive or critico-justificational, makes use of implicit *a priori* categories. Therefore the subject who makes justificational and argumentative inferences regarding his own beliefs may not, or cannot, act in full awareness. Every argument presupposes or implies some tacit *a priori* categories – or meta-conscious categories, as Boudon prefers to say, following F. Von Hayek (see Von Hayek, 1973). This character is given to them also, Boudon underlines, from the fact that the *a priori* categories are accepted uncritically, as they often have a widely accepted validity.

> En un mot, le fait que le sujet connaissant soit condamné à utiliser de façon non critique des *a priori* très généralement valides est l'un des nerfs de la connaissance et, en même temps et de façon indistincte, une source essentielle de ses erreurs et de ses croyances en des idées douteuses, fragiles ou fausses (Boudon 1990, p. 110).

If *a priori* categories represent an indispensable ingredient in all explanations, or in all judgements or arguments, it follows that, to the extent that implicit statements can be made explicit, reasoning that once seemed linear will come to seem circular. For example, suppose we have a theory that includes the following explicit statements:

1) p
2) $p \rightarrow q$
3) $q \rightarrow r$

---

4) r

If this becomes integrated with implicit propositions 5, 6, and 7, it becomes:

5) r
6) $r \rightarrow p$
7) and 1) p
2) $p \rightarrow q$
3) $q \rightarrow r$

---

4) r

> Or la théorie explicite constituée par les propositions 1) à 4) est *linéaire*. La théorie incluant de surcroît les propositions 5) à 7) est, quant à elle, *circulaire* (Boudon 1990, p. 112).

This model, in Boudon's judgement, can also clarify the nature of inductive reasoning, which has so troubled epistemologists in past years (at least from Hume to Popper), because it reveals its circular character. Boudon uses an example taken from Peirce:

> Tous les haricots de ce sac ont une couleur identique.
> *Quelques* haricots de ce sac sont blancs.
>
> ---
>
> *Tous* les haricots du sac sont blancs.

With respect to form, the inference is faultless. The problem is to know if the first proposition is valid. To solve it, we must begin with particular observations that, to the extent to which they are convergent, reinforce the conjecture of the proposition. Thus, when the argument is made explicit, it will take the following form:

1 *Jusqu'ici*, tous les haricots que j'ai tirés de ce sac sont blancs.
2 [(Cela suggère la *conjecture* que) tous les haricots du sac ont une couleur identique].
3 Les haricots de ce sac ont tous la même couleur.

---

4 *Tous* les haricots du sac sont blancs.

This argument is a circular one, therefore not valid. However, most often its non-validity is not detected, because we pay attention only to its explicit part without integrating it with the implicit part, and in so doing, we fail to notice that premise 3 is based on statements 1 and 2. In other words, inductive reasoning presupposes implicit, even unconscious statements, and as Boudon makes clear, it is just this part that gives confidence *('confiance')* to the conclusion. In a certain sense, the faith that one assigns to the reasoning serves to hide the implicit proposition that is in fact a conjecture, and that therefore invalidates the reasoning. Implicit propositions are perceived meta-consciously, and the subject trusts them:

> Les *conjectures* que le sujet connaissant émet constamment dans la vie quotidienne comme dans la recherche scientifique ne sont pas *vécues* ou *perçues* comme telles, mais sont plutôt vues par lui comme des évidences, en tout cas comme des propositions que lui paraissent à ce point aller de soi qu'il ne songe pas à les mettre en doute (Boudon 1990, p. 118).

The model explains how the subject can judge subjectively that certain reasons, while being objectively wrong, are good reasons. At the level of common sense, we might easily concede that an individual may have had 'good reasons' for behaving as he did in particular circumstances, even if he was wrong. To overcome this contradiction, Boudon formulates the concept of subjective rationality (a concept close to that of *bounded rationality* put forward by H. Simon – see Simon 1982). Reasons are narrow and incomplete because the cognitive resources immediately available to the subject, who acts in particular circumstances, are incomplete. Still, the subject's reasons do not lack the distinctive properties of objectively valid rationality. Indeed, reason can make conjectures based upon sound principles *('principes solides')*, and these principles are sound to the subject, because he perceives them as evident and reliable.

This outcome of Boudon's thought indicates that 'reasons' are not constituted, so to speak, solely of *reason*, but are also intrinsically linked to other qualities that are largely non-cognitive and non-rational, such as trust, sentiments and beliefs. Now, if *'"la rationalité subjective" est le produit normal de la discordance entre la complexité des situations devant lesquelles le sujet se trouve placé et le caractère limité de ses capacités cognitives'* (Boudon, 1990, p. 377), then we must acknowledge that the search for 'good reasons' alone leads to an incomplete explanation of action. Action always implies, on the part of the subject, the putting into act of certain theories, conjectures, principles, or the *a priori*, together with other, no less important, elements and qualities, which are not reducible *in toto* to these.

If, to the subject, they did not seem evident – if he did not trust the premises of the conjectures – would he realise that his reasons are not 'good reasons' for doing what he does, or for believing what he believes? How could that which is not evident to him appear evident? How could he trust principles that are not objectively founded? If subjective reason were not limited, could one maintain that the acting subject could make a mistake? Is the explanation given by Boudon perhaps a circular one? To me it seems that, taking account of the due differences, the explanation offered by Boudon is not greatly different from that offered by Lévi-Bruhl, which Boudon nevertheless criticises. These problems derive from the fact that Boudon has abandoned the Weberian model of social action and of social relationships, and also, naturally, from his own concept of complex action which he formulated in his work *La logique du social* (cf. Boudon 1979).

6

Boudon criticises Lévi-Bruhl's explanation of magic beliefs, and his critique represents an important point in his cognitive theory of rationality. Lévi-Bruhl, argues Boudon, explains the beliefs of primitive societies by assuming that the mental constitution of primitive peoples differs from that of western peoples. For example, in primitive societies, the medicine man 'confuses verbal similarities and real similarities, and, in general, the relations between words and the relations between things' (Boudon 1995-1997, p. 106), because he has a different mental constitution. The renowned theory of the 'primitive mentality', in fact, analyses beliefs as 'the effect' of unconscious causes, beyond the individual's control. In Boudon's previous work, *L'Art de se Persuader*, Boudon maintains that Lévi-Bruhl conceives the theory of primitive mentality as the effect of primitive culture.

From the outset, Boudon asserts that Lévi-Bruhl's explanation is circular, in that his explanatory cause is assumed in the effects that it should explain, whereas a causal explanation is necessarily linear. Lévi-Bruhl's mistake consists in his having held that the beliefs of primitive peoples, given that they appear irrational to our mentality, are the effect of 'a mental constitution' that differs from ours; that they are, then, the product of a particular, primitive, mentality.

By contrast, if Lévi-Bruhl's starting point had been from the principle that primitive peoples have 'good reasons' for their beliefs, he would have been compelled to recognise that these good reasons are the product of the same constitution as ours. In fact, Lévi-Bruhl does not think that 'good reasons' are irrational, but that they correspond to a mental constitution that he deems pre-logical. In a certain sense, he makes the same mistake as Pareto does, when the latter distinguishes between logical and illogical social actions. Lévi-Bruhl is well aware that the mental processes that regulate actions, hypotheses and beliefs are not uniquely reducible to the formal aspect of mental processes, since they relate to the sentiments, passions, knowledge and experience of individuals. These are complex processes, Boudon would say.

Lévi-Bruhl does not question the existence of magic beliefs and the reasons of individuals. As a philosopher, he is greatly interested in tracing the mental or cognitive processes that regulate the formation of beliefs or categories which shape the experience of primitive peoples. He holds that these beliefs are a constitutive part of their social experience, and of the way in which they see and explain the world. It is this aspect that Boudon overlooks, although we shall not be examining it further here. According to Boudon, Lévi-Bruhl's error is in holding that the beliefs of primitive peoples are caused in an unconscious or unknowable way by their pre-logical 'mental constitution'. This conviction pushes him towards thinking that the activities of primitive man and of civilised man are controlled by different cognitive and mental forces, in that the processes of primitive man present some aspects that do not correspond to the logical forms that underlie the activities of civilised man. In other words, Lévi-Bruhl does not maintain that the mental activities of primitive man are similar to those of civilised man, and that, consequently, primitive peoples have 'good reasons' for holding their beliefs, which, according to the methodological principle defended by Boudon, constitutes the most economical means of explanation. Nevertheless, we must not overlook the fact that Lévi-Bruhl attempts to explain beliefs by referring to the cognitive activities that underlie them. He thinks that they are unconscious, says Boudon, when instead they would have to be conscious in certain ways.

In his preceding work, *L'Art de se Persuader*, Boudon classifies Lévi-Bruhl's explanation as one of type IIb. This type covers the explanations that cannot be identified with reasons or passions ('the heart has reasons that reason does not know', as says Pascal), but instead are identified with a particular 'mental organisation'. This type of explanation does not make use of Kant's cognitive structure of the *a priori*. Collective representations, such as beliefs presuppose mental operations that differ from those that underlie the elaboration of Newtonian science, as Kant had exemplified. Following Emile Durkheim, collective representations correspond to the way in which individuals in a given society take part in and share their social experiences. It is because they live together that the individuals have shared experiences, and elaborate them into collective representations. These are collective representations because they represent the common experiences of individuals, and these would not exist if the individuals were isolated from one another. The individuals that live a common experience form representations, the characteristics of which cannot be attributed *in toto* to the representational capacities of the individual (although they are connected to individual faculties). This hypothesis, or conjecture, directs Lévi-Bruhl's research.

It does not seem correct to me, as it does to Boudon (and this is the second aspect of his critique), to say that Lévi-Bruhl sets *'les mécanismes logiques et psychologiques [...] comme variables selon les cultures et les sociétés, et par conséquent comme étant tributaires de causes sociales'* (Boudon 1990, p. 25). This type of explanation is considered by Boudon to be an example of cultural relativism (cf. Needham 1972), a position which he criticises and refutes.

Indeed, Lévi-Bruhl conjectures that there exists a correspondence between representation, thought, and social reality, and so he assumes that there are many perspectives – many ways to see the world:

> [We must] [...] *renoncer à ramener d'avantage les opérations mentales à un type unique, quelles que soient les sociétés considérées, et à expliquer toutes les représentations collectives par un mécanisme psychologique et logique toujours le même* (Lévi-Bruhl 1960 [1922], pp. 19-20).

In our judgement, as outlined above, Lévi-Bruhl is perfectly conscious that Kant's *gnoseology* is inadequate for comprehending and explaining social phenomena, especially those societies that differ markedly from ours. This point is repeated once again in another excerpt from Lévi-Bruhl quoted by Boudon:

> Dans les représentations collectives de la mentalité primitive, les objets, les êtres, les phénomènes peuvent être, d'une façon incompréhensible pour nous, à la fois eux-mêmes et autre chose qu'eux-mêmes. D'une façon non moins incompréhensible, ils émettent et ils reçoivent des forces, des vertus, des qualités, des actions mystiques qui se font sentir hors d'eux, sans cesser d'être où ils sont (Lévi-Bruhl 1960 [1922], p. 77).

Lévi-Bruhl stays faithful to this definition in all of his work. Although he later changes his opinion on the notion of pre-logical thought, he does not give up the idea that the primitive way of thinking (mysticism and participation) does not belong to the sphere of reason. Yet he does not fall victim to obscurity and irrationalism. Essentially, the experience of primitive people is emotional:

> Or quand on parle de participation, il s'agit d'une consubstantialité, d'une communion, d'une identité même (identité-dualité) entre des êtres et des objets. Ce qui n'est pas représenté, en général, c'est cette relation intime, si caractéristique, entre ces êtres et ces objets. Elle est sentie; mais les objets eux-mêmes sont représentés (Lévi-Bruhl 1949, p. 139).

E. Lévinas strongly emphasises Lévi-Bruhl's effort to comprehend 'primitive mentality', and on this point quotes a passage from Lévi-Bruhl's last work:

> [One should not take it for granted] [...] que les êtres sont *donnés* d'abord, et ensuite entrent dans des participations. Pour qu'ils soient données, pour qu'ils existent, il faut déjà des participations. Une participation n'est pas seulement une fusion, mystérieuse et inexplicable, d'êtres qui perdent et conservent à la fois leur identité [...] Sans participation, ils ne seraient pas *données* dans leur expérience: ils n'existeraient pas. [For the individual, participation] est une *condition de son existence*, peut-être la plus importante, la plus essentielle pour cette mentalité exister c'est participer à une force, à une essence, à une réalité mystiques (Lévi-Bruhl 1949, pp. 250-251).

I have discussed at length Boudon's critique of Lévi-Bruhl for several reasons:

1 Lévi-Bruhl offers an explanation of the beliefs of primitive peoples that is centred fundamentally on their 'mental constitution', holding that these people make use of procedures of justification that differ from those used by civilised men.
2 Boudon, in his turn, looks for an explanation of beliefs centred almost exclusively on the mental operations – the reasoning – which underlies them.
3 Boudon devises, as does Lévi-Bruhl, a cognitive model that explains how it is possible to have false beliefs.
4 Boudon, unlike Lévi-Bruhl, retains that an individual, independently of others, can have good reasons for holding beliefs, and that, consequently, if more individuals share the same beliefs, this is due to the fact that they logically justify them in the same way, and not to social causes such as interiorisation, socialisation, interdependence, the structural system, and culture. Instead, beliefs depend on cognitive causes. Lévi-Bruhl, rather, thinks that individuals share beliefs because they share them in their social experiences. Further, Lévi-Bruhl thinks that there exists a correspondence between social experience and mentality.

7

In Boudon's view, a sociological theory must include a general theory of action, and must satisfy the principle that it is not possible to understand the actions of social actors if we do not know the system of interactions to which they belong. This precept does not at all indicate that the actors' behaviour is determined solely by the system of interactions, as each actor tries to make the most convenient decisions according to his or her perceived interests (primacy of action). However, Boudon does not clarify how it is possible to know the interaction system in which the actor places his or her actions; nor does he clarify how the actor, through his or her actions, contributes to building the interaction system in which he or she is involved, given that an interaction system does not simply result from the aggregation of, or from the actions of, the individual actors who are more or less directly involved in it.

A sociologist wishing to understand an actor's choices must know the conditions in which that actor operates – but how to know such a situation? The actor's choice also concerns the situation in which he or she acts; the choice, then, is such because it is considered adequate to the situation. The action must take place within a situation or conditions that make it possible, defining, for the actor, the set of possibilities for his or her action, without thus having to sustain that the action is determined by the interaction system in which it is necessarily carried out.

Now, the fact that an individual is involved in an interaction system gives him or her certain properties that he or she could not otherwise have had (although it is difficult to imagine an individual who doesn't belong to any interaction system). Such properties are derived from his or her belonging to the system. They are relational properties, not intrinsic to the subject, even if the subject enacts them.

Let us take an example given by Boudon himself: that of the queue that forms at the local bakery on Sunday morning. If I want to buy pastries on Sunday morning, I cannot avoid the annoyance of having to queue at the shop, because other people will have decided, as I have, to buy pastries. (Indeed, it is a tradition, in France and elsewhere, to have a sweet as part of Sunday lunch.)

A queue presupposes an order that is, in general, respected by everybody, and the person at the head of the queue is served first, followed by each of the others in turn. I, along with many others, do not wish to give up this tradition, in spite of my having to stand in the queue and wait my turn (although, on the other hand, waiting in a queue can be an occasion to chat with acquaintances or to exchange information, and so on). My action depends on other peoples' actions in the taking of turns and respecting the order. Finally, my action depends not only on *my* desire, but also on the desires of those who are waiting for me to bring home the cakes.

The baker, for his part, acts in order to meet the demands of his Sunday customers, and also in accordance with the tradition; if he does not employ extra shop assistants in order to minimise the queue, we must assume that this is because of his own interests.

If, in Boudon's interpretation, the queue at the bakery is an effect that is undesired by individual actors, it is also one that is caused by a tradition that individuals accept, and on the basis of which they conform in their behaviour. We might say that the queue is an effect of the behaviour of individuals, yet it exhibits its own internal order, which imposes itself on those individuals – while tradition is something that individuals have received and that they accept to such an extent that it pushes them to behave as they do.

The queue is produced by customers who have 'good reasons' for standing in the queue; their behaviour is adequate to axiological rationality: that is, to the value that this particular tradition has perpetuated (M. Weber). However, how can we not recognise that these 'reasons' have to do with tradition, which the actors have made their own through sharing it as part of their being members of the same culture? The 'good reasons' that customers have for buying pastries on Sunday morning are not sufficient to explain this tradition, while the acceptance of the tradition contributes to explaining the formation of the queue. We may ask whether it is the queue, as an ordered system, which brings individuals to conform in their behaviour, or the cognitive rationality of each individual (that is, the 'good reasons' that everybody has for behaving in a certain way), that drives each person to wait his or her turn in the queue. The later Boudon leans toward the second alternative, but in this case he abandons his concept of complex action.

Boudon inadvertently moves the discourse from the analysis of interaction systems and of reasons and beliefs, to the analysis, alone, of the 'normal functioning of thought' through which these reasons and beliefs are deducible (cf. Boudon 1995). He considers this analysis to be sufficient for understanding the phenomena that actions create. This move involves, in his latest work, the abandonment of social interaction, which, by contrast, is the very hub of the sociological analyses of Simmel and Weber. The fact that only an individual can have reasons and beliefs does not exclude the additional fact that reasons and beliefs are shared, and that individuals elaborate them and accept them in their interactions.

8

Critics, in general, have not paid attention to the fact that for Weber the meaning of social action is realised in relation to the action of more individuals, be they simultaneously present or simultaneously absent. Man, according to neo-Kantian philosophy to which the German sociologist adheres, and within which he frames his epistemological considerations, is undoubtedly a *Sinngeber*; but the object of sociology is social action, whose meaning, for the actors, lies in the relation and in accordance with the expectations and orientations of the individual actors who are involved in their various individual ways:

> Soziale 'Beziehung' soll ein seinem Sinngehalt nach aufeinander gegenseitig *eingestelltes* und dadurch orientiertes Siechverhalten mehrerer heiβen. Die soziale Beziehung *besteht* also durchaus und ganz ausschlieβlich: In der *Chance*, dass in einer (sinnhart) abgebbaren Art sozial gehandelt wird, einerlei zunδchst: worauf diese *Chance* beruht. (Weber 1972, p. 13)

In order that one may speak of social relationships, certain requirements must be fulfilled, as exemplified by Weber himself:

a) there must be 'a minimum of *reciprocal* relationship in the agency *of both parties*' ('ein Mindestmaß von Beziehung des *beider* seitigen Handelns *aufeinander*');
b) there must be a content of sense that allows the parties to enter into relationships with each other – this 'can be very varied: combat, enmity, sexual love, friendship, reverence, trade exchange' ('Kann der allerverschiedenste sein: Kampf, Feindschaft, Geschlechtsliebe,

Freundschaft, Pietät, Marktaustausch') (Weber makes explicit reference to relational terms);

c) in 'social formations', such as 'the state', 'the church', 'the company', 'marriage', etc., 'the social relationship [consists] exclusively and simply in the *possibility* that a determined reciprocal agency has taken place, or will take place, according to its content of sense' ('Die soziale Beziehung, *besteht*, auch wenn es sich um sogenannte [...] ausschließlich und lediglich in der *Chance*, daß ein seinem Sinngehalt nach in angebbarer Art aufeinander eingestelltes Handeln stattfand, stattfindet oder stattfinden wird') – that is, according to the sense given by the fact that the social relationships are between spouses, friends, companions, etc. – and any such formations 'cease to "exist" *sociologically* as soon as the possibility disappears of carrying out some specific forms of social action orientated on the basis of sense' – in other words, as long as there are those specific 'forms of social action', there will also be the content of sense of social relationships;

d) with those who participate in 'reciprocally established actions', it must not be said that they attribute to the relationship the same 'content of sense', nor that they have the same internal disposition towards one another and that they share the same meaning – but also, in this case, a reciprocal reference to one another is present;

e) a relationship is to be considered bilateral when the content of sense of one of two actors corresponds to that of the other – for example, father and son;

f) a social relationship can be transitory or long-lasting;

g) the content of sense of a social relation may change;

h) the permanent content of sense of a relation can be 'formulated as "precepts"';

i) the content of sense can be *stipulated* through reciprocal commitment.

In order to have a social relationship, the possibility must exist of acting according to a determined disposition in a way which conforms to a sense that is *'intentioned on average'* – that is, shared by the acting parties. A social relationship is established in a manner such that among social parties the possibility is realised of the '*recurrence* of a certain attitude corresponding to the sense' (that is, valid in relation to it and expected in conformity to it). The content of sense constitutes the social relationship. Therefore a social relationship is such only inasmuch as it is realised, is possible, and is set according to a sense intentioned 'on average', intersubjectively; which is to say that a social act is always based upon the possibility of existing in a specific way endowed with sense 'on average'. At the same time, it indicates the possibility of it, as it makes possible actions endowed with sense, and it also refers to other possibilities, so that only among actions of different individuals are actions endowed with sense established.

The reference to Weber's position is aimed at underlining how a theory of social action cannot leave out of consideration the analysis of the constitution of 'content of sense' for social actors, which represents the *medium* not only for social interaction but also for the actions of an individual. As a consequence, the rationality of action cannot be reduced to the 'normal functioning of thought', as this is measured with different situations, and, in turn, is the result of the numerous situations in which it is exercised. It can be dangerous to speak of a 'normal functioning of thought' independently of the many and various situations in which it is applied, because it is present in the different objective structures of actions.

Bibliography

R. Boudon, *La crise de la sociologie*, Librairie Droz, Genève, Paris 1971.

R. Boudon, *La logique du social*, Librairie Hachette, Paris 1979.

R. Boudon, F. Bourricaud, *Dictionnaire critique de la Sociologie*, Puf, Paris 1982.

R. Boudon, *L'idéologie, ou l'origine des idées reçues*, Fayard, Paris 1986.

R. Boudon, *L'art de se persuader des idées fausses, fragiles et douteuses*, Fayard, Paris 1990.

R. Boudon (ed.), *Traité de sociologie*, Puf, Paris, 1992.

R. Boudon, 'La logique des sentiments moraux', *L'Année sociologique*, n. 44, 1994.

R. Boudon, *Le juste et le vrai. Études sur l'objectivité des valeurs et de la connaissance*, Fayard, Paris 1995.

R. Boudon, A. Bouvier, F. Chazel (eds.), *Cognition et science sociales*, Puf, Paris 1997.

A. Bruschi, *La teoria dei modelli nelle scienze sociali*, il Mulino, Bologna, 1971.

E.M. Classen (ed.), *Des fondements philosophiques des systèmes économiques*, Payot, Paris 1967.

L.A. Gérard-Varet, J.C. Passeron (eds.), *Calculer et raisonner: les usages du principe de rationalité dans les sciences sociales*, EHESS, Paris 1993.

G.G. Granger, 'Les trois formes de la rationalité économique', in L.A. Gérard-Varet, J.C. Passeron (eds.), *Op. cit*, 1993.

F.A. von Hayek, *Law, Legislation and Liberty: a New Statement of the Liberal Principles of Justice and Political Economy*, Routledge and Kegan Paul, London, 1973.

L. Lévi-Bruhl, *La Mentalité primitive*, PUF, Paris, 1960, (1st ed., 1922).

L. Lévi-Bruhl, *Les carnets*, PUF, Paris 1949.

E. Lévinas, 'Lévi-Bruhl et la philosophie contemporaine', in *Revue philosophique de la France et de l'étranger*, n. 4, 1957.

R. Needham, *Belief, Language and Experience*, Basil Blackwell, Oxford 1972.

K. Popper, 'La rationalité et le statut du principe de rationalité', in E.M. Classen (ed.), *Op. cit.*, 1967.

R. Rorty, *Philosophy and the Mirror of Nature*, Basil Blackwell, Oxford 1980.

J.R. Searle, *Rationality in Action*, MIT, New York, 2001.

H. Simon, *Models of Bounded Rationality*, MIT Press, Cambridge 1982.

M. Weber, *Wirtschaft und Gesellschaft*, Fünfte, Revidierte Auflage, Besorgt Von Johannes Winckelmann, J.C.B. Mohr (Paul Siebeck), Tübingen 1972, It. transl., P. Rossi, *Trattato di sociologia*, Il Mulino, Bologna 1996.

Giuseppe Padovani, Dipartimento di Studi Politici e Sociali, Università di Parma

Roberto Poli

Ontological Models

Abstract:

After distinguishing three different types of ontology, the connection between logic and ontology is discussed. The thesis is defended that formal complexity should be distinguished from ontological complexity. The thesis is tested against the phenomenon of duality and in particular against the distinction between states of affairs, situations of affairs and unfolded states of affairs. The Non-Fregean Logic developed by the late Suszko is used to provide a few hints about one of the possible strategies that may be used to formally substantiate those distinctions.

Keywords:
duality, ontological levels, state of affairs, decomposition, parameterization

1 Descriptive, formal and formalized ontologies

The unity and the variety of the world is the outcome of the complex interweaving of dependence connections and forms of independence among the many items of which it is composed. I shall seek to explain the features of this multiplicity by beginning with an apparently trivial question: what is there in the world?

We may say that there are material things, plants and animals, as well as the products of the talents and activities of animals and humans in the world. This first almost trivial list already indicates that the world comprises not only things, animate or inanimate, but also activities and processes and the products that derive from them.

It is likewise difficult to deny that there are thoughts, sensations and decisions, and the entire spectrum of mental activities. Just as one is compelled to admit that there are laws and rules, languages, societies and customs.

We can set about organizing this list of objects by saying that there are *independent items* that may be *real* (mountains, flowers, animals, and tables), or *ideal* (sets, propositions, values), and *dependent items* which in turn may be *real* (colours, kisses, handshakes and falls) or *ideal* (formal properties and relations).

All these are in various respects items of the world. Some of them are actually exemplified in the world in which we live; others have been exemplified in the past; and yet others will possibly be exemplified in the future (Poli 2001, ch. 5).

Descriptive ontology concerns the collection of such *prima facie* information either in some specific domain of analysis or in general.

Formal ontology distills, filters, codifies and organizes the results of descriptive ontology (in either its local or global setting). According to this interpretation, formal ontology is formal in the sense used by Husserl in his *Logical Investigations* (Poli 1993). Being 'formal' in such a sense therefore means dealing with categories like *thing, process, matter, whole, part, and number*. These are pure categories that characterize aspects or types of reality and still *have nothing to do with the use of any specific formalism*.

Formal codification in the strict sense is undertaken at the third level of theory construction: namely that of *formalized ontology*. The task here is to find the proper formal codification for the constructs descriptively acquired and formally purified in the way just indicated. The level of formalized constructions also relates to evaluation of the adequacy (expressive, computational, cognitive) of the various formalisms, and to the problem of their reciprocal translations.

The close similarity between the terms 'formal' and 'formalized' is somewhat unfortunate. One way to avoid the clash is to use 'categorical' instead of 'formal'.

Most contemporary theory recognizes only two levels of work and often merges the level of the formal categories either with that of descriptive or with that of formalized analysis. As a consequence, the specific relevance of categorical analyses is too often neglected.

The three levels of ontology are different but not separate. In many respects they affect each other. Descriptive findings may bear on formal categories; formalized outcomes may bear on their twin levels, etc. To set out the differences and the connections between the various ontological facets precisely is a most delicate task (Poli 2002).

2 The logical and the ontological

We have seen that ontology comes in various guises. Logic itself presents a number of different facets. In this regard, the following quotation from Gödel is exemplary in its clarity:

> Mathematical logic [...] has two quite different aspects. On the one hand, it is a section of Mathematics treating of classes, relations, combinations of symbols, etc. instead of numbers, functions, geometric figures, etc. On the other hand, it is a science prior to all others, which contains the ideas and principles underlying all sciences (Gödel 1944, p. 123).

Gödel's description of mathematical logic as the science that precedes all others is entirely similar to the definition of metaphysics provided by Aristotle. It is accordingly natural to think that mathematical logic (in Gödel's second sense) is able to depict the deepest-lying structural maps of reality. The two aspects of mathematical logic distinguished by Gödel can be respectively denoted with the expressions, taken from Heijenort 1967, of 'logic as calculus' and 'logic as language' (cf. Cocchiarella 1974 and 1988).

Their difference resides in that logical form can be used to represent many different things: logical validity, truth conditions, an abstract calculus, cognitive aspects or even *ontological structures*. Gödel's reading suggests that logic (as language) is formal ontology.

The thesis of the identity of logic (as language) with (categorical) ontology is, as we know, one of the essential assumptions of Platonism (or logical realism). Moreover, it is one of the fundamental theses of the Hegelian dialectic and its paralogisms.

Unfortunately, the connections between the ontological and formal levels are much more complicated than would appear from Gödel's observation. I shall therefore resist Gödel's solicitations and refrain from identifying logic (as language) *sic et simpliciter* with ontology.

Let us assume that categorical ontology and the formal sciences both have the task of searching for what is universal, in general or in some domain. The problem is that we have no reason to think that what is universal in the formal sense must perforce coincide with what is universal in the ontological sense. In other words, it is advisable to keep what is universal in the formal sense separate from what is universal in the ontological sense (unless there is some strong argument to the contrary).

One way to view the difference between the two accounts of universality is as follows. Let us imagine that the structures of universality concretize themselves in simplicity, in accordance with the ancient motto *simplex sigillum veri* ('simplicity is the seal of truth'), so that the more universal structures are also simpler (according to some criterion of simplicity). If this were true, ontologically simpler (and therefore more universal) structures should also be formally simpler (and vice versa).

Perhaps the most elementary example is the one that follows. Consider the universe of states of affairs and the propositions that refer to them. As we know, when adequately rigid ontological conditions are imposed upon states of affairs (i.e. upon the objects of the universe of discourse), the usual propositional connectives are interdefinable. If we instead relax the ontological conditions and therefore characterize the universe of discourse in a more universal way, connectives are no longer interdefinable (as in intuitionist logic).

Greater ontological rigidity may therefore simplify the formal level, but greater ontological universality may make it more complex.

3 Duality

In the light of these considerations we may assume that it is possible to distinguish logic (as language) from ontology. Having said this, however, perhaps the most interesting aspect of the relationships between logic and ontology concerns neither their unification nor their distinction but their *correlation*, or in other words, their reciprocal influence. An area of particular significance for the study of the correlations between logic and ontology is that of duality phenomena (in both the formal sciences and ontology).

Duality phenomena are well known in the formal sciences: suffice it to mention situations like ideal/filter, open set / closed set, join/meet, intersection/union, universal quantifier / existential quantifier, it-is-necessary-that/it-is-possible-that. While analyzing the numerous cases of formal duality, Rosado Haddock (1991) points out that (1) the feature shared by duals is their interderivability within a particular theory, and that (2) interderivable propositions can be considered to be different states of affairs which relate to the same situation of affairs. Rosado Haddock thus takes up Husserl's distinction between *Sachverhalt* and *Sachlage*. David W. Smith (1991) objects to the strictly formal criterion used by Rosado Haddock on the ground that Husserl's distinction between states of affairs and situations of affairs is not essentially formal but mainly ontological.

Strictly speaking, however, both Rosado Haddock and Smith are right, because duality is a phenomenon with both formal and ontological aspects. In what follows we shall also see that the discussion between Rosado Haddock and Smith has failed to address what is perhaps the most interesting aspect of duality: the presence of two sharply distinct families of dual phenomena.

4 States of affairs and situations of affairs

The following example is taken from Aristotle and was quoted by Anscombe and Geach 1961:

> The road that leads from Athens to Thebes is the same road that leads from Thebes to Athens: but in the former case it goes uphill, while in the latter it goes downhill.

We may therefore say that this road, *qua* road from Athens to Thebes, is uphill, and that this same road, *qua* road from Thebes to Athens, is downhill. One thus concludes that there is a relation between the object and the standpoint or the point of view of looking at it. In entirely similar terms, Anscombe states that, 'there aren't such objects as an A *qua* B, though an A may, *qua* B, receive such-and-such a salary and, *qua* C, such-and-such a salary' (Anscombe 1981b, 208; the role and the properties of the functor *qua* are analyzed in Poli 1994 and 1998).

It is well known that, for Husserl, cases like 'A is part of B', 'B contains A as its part', are two *states of affairs*, which have the same *situation of affairs* as their foundation. Corresponding to the same situation of affairs may be two or more states of affairs; in the same way as corresponding to the same state of affairs may be two or more propositions (thoughts) (Rosado Haddock 1991).

There follow examples of states of affairs that refer to the same situation of affairs (Smith 1991):

- Heidi is Hans's wife; Hans is Heidi's husband;
- The glass is half full; the glass is half empty.

Husserl constructs states of affairs from situations of affairs on the basis of the following principles:

- States of affairs are categorical structures;
- Situations of affairs are precategorial entities;
- States of affairs are founded upon situations of affairs;
- Different states of affairs may be founded upon the same situation (Smith 1991, p. 52).

While commenting on these principles, David W. Smith adds:

> Unfortunately, the principles Husserl put forth do not tell us what situations are. Indeed, some four different intuitions might lie behind these principles, yielding four different notions of situation, outlined as follows:
> 1. […] Situations are just *species*, or types, of states of affairs […]
> 2. States of affairs might be thought of as *aspects* of situations (in effect reversing the first proposal) […]
> 3. A situation might be viewed as the matter *(Sache)* from which different states of affairs are formed […]
> 4. A situation might be viewed as a certain sort of *part-whole* complex, or affair *(Sache)*, from which parts are extracted and put together into states of affairs […]
>
> What remains unclear, then, is the kind of unification that form a situation (Smith 1991, pp. 53-54).

The following table summarizes the situation:

Table 1

	Situation of affairs	State of affair
	Species	Instance
	Instance	Aspect
	Matter	Form
	Whole	Part

In reality none of these oppositions seems to be correct. Consider the difference between 'The road that leads from Athens to Thebes' and 'The road that leads from Thebes to Athens'. Their difference seems manageable between descriptions based on 'pure' or 'a-directed' relations as opposed to descriptions based on 'directed' relations. Consider the difference between the following directed and a-directed diagrams:

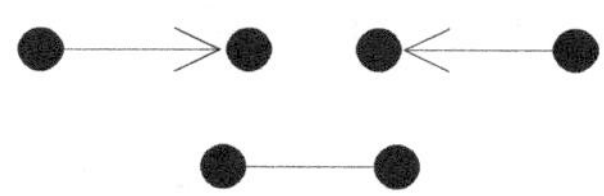

The two directed graphs in the first row can be interpreted as 'the road that leads from Thebes to Athens' and 'the road that leads from Athens to Thebes'. These interpretations are explicitly linked to a direction. On the contrary, the graph drawn below 'says' *only* that the two vertices (points) are connected. In other words, it exemplifies only the pure situation of 'being connencted to'. It is clear that the graphs in the first row result from the graph drawn in the one below as soon as an indication of direction is added (and vice versa).

We may therefore distinguish between two representational spaces: a space composed of situations of affairs (a-directed graphs), and a space composed of states of affairs (directed graphs).

If this reading of the difference between states of affairs and situations of affairs can be generalized, then the relationship between situations and states would have the nature of a theory-theory connection and would not correspond to any of the four oppositions in Table 1.

5 Unfolded state of affairs

The situation can be further elaborated by introducing a third representational space that mirrors the difference between the ontological viewpoint and the judgements. Daubert first discussed this opposition. In the words of Karl Schuhmann and Barry Smith:

> To capture this opposition between how things stand in themselves and how they are asserted to be in our judgements, Daubert [...] distinguishes between the *Sachverhalt* and what he calls *'Erkenntnisverhalt'*, the 'state of affairs as cognized' or as 'unfolded' in cognition [17v]. The former is the objectively existing structure of things, properties and relations as they are in and of themselves. The latter is that side or aspects of the former which serves as the immediate objectual correlate of a given concrete act of judging [...]
> Thus consider: 'The chairman opens the meeting', 'The chairman is opening the meeting', 'The meeting is being opened by the chairman', 'The opening of the meeting is being conducted by the chairman', 'The chairman has opened the meeting', 'The meeting has been opened by the chairman' [17v, 63r]. Each of these sentences differs as to its associated state of affairs as cognized, but they are in fact concerned with one and the same objective *Sachverhalt* (Schuhmann and Smith 1987, pp. 367-8).

The examples quoted require a theory able to unify them into a coherent whole. This will contain 'chairman', 'meeting', 'opening the meeting', and so forth, as primitive or derived terms. It will be about the worldly structure of meetings – that is, it will be about how things are.

It follows that at least three different levels can be distinguished: the level of the *Sachlage*, the level of the *Sachverhalt*, and the level of the *Erkenntnisverhalt*. From a formal point of view, aspects of the latter can be represented by allowing more edges between two adjacent vertices (as in multigraphs).

Let us now try to construe the logical table corresponding to the ontologically-grounded distinctions so far considered.

Table 2

Ontological	Logical
Erkenntnisverhalt	*Utterance*
Sachverhalt	*Sentence*
Sachlage	*Situation*
(Evidence)	*Truth-value*

Both sides present a many-one form of connection. As regards the logical side, this means that many utterances may express one sentence; many sentences may express one situation; and many situations may express one truth-value. By resorting to other conceptual frameworks, we may also say that utterances instantiate judgments; sentences instantiate senses (categorical objectualities); situations instantiate things-in-themselves (pre-categorical objectualities); and truth-values instantiate [...] well, nothing but themselves.

In their turn, all of these are values of a common, basic argument. Call it 'proposition'. We will say that propositions have utterances, sentences, situations and truth-values as their functional values. What this means will soon become clearer.

The ontological side presents the very same many/one structure as the logical one. Many unfolded states of affairs may refer to one state of affairs, etc. The ontological rock-bottom level represented by 'evidence' has been bracketed because used in this way it is a more Brentanian than Husserlian concept. All the ontological categories are values of a common argument which may be called 'thought' or 'content'.

Let me summarize the discussion thus far. On analysing a number of classic situations of categorial duality, we have seen the possibility of an underlying precategorial unification (the relationship between states of affairs and situations of affairs). The unification therefore acts as a quotient or invariant. I then generalized the procedure by unifying a multiplicity of unfolded states of affairs in their underlying state of affairs. The two steps were made uniform by constructing a table of quotients in which multiplicities of items were progressively unified at increasingly profound levels. The table of quotients was also constructed in two different guises, one formal and one ontological.

If we operate in this manner, duality phenomena become a particular case of a more widespread phenomenon. On this basis, which we may take as given, we can now deepen our examination of categorial dualities.

6 Ontological duality

The importance of duality emerges as soon as one understands that there are two significantly different families of dual phenomena distinguished by the result obtained when iterating the duality operation. Given an item C, we may use C* to denote the dual of C. The fundamental difference between the two families of dualities thus becomes the difference between the case $C^{**} = C$ and the case $C^{**} \neq C$. I shall call the first form of duality an 'elementary duality' and its complement a 'non-elementary duality'.

Non-elementary duality is a fundamental aspect of the categorical analysis of ontology. For the sake of simplicity, here the term 'categories' refers to both principles and categories in the strict sense (on the difference between categories and principles in Aristotle see Poli 2001, p. 59).

To show how broad is the phenomenon of categorical duality, suffice it to mention some of the best-known categorical pairs.

Besides pairs of categories with a recognized mathematical meaning, like finite/infinite and discrete/continuous, one easily finds many further categorical pairs that have figured prominently in the history of philosophy. For example:

- matter/form
- potential/act
- quality/quantity
- one/many
- identity/difference
- individual/universal
- part/whole

A moment's reflection shows that the dualities that hold within these pairs of categories *are not elementary*: the form of a certain substance is the substance of another form different from the initial one; the act of a certain potential is the potential of a new act different from the initial one; and so on.

The categorical duality that holds in ontology (at least with regard to the pairs of categories listed above) is therefore not an elementary duality.

7 A few distinctions

The final part of this article analyzes (some of) the distinctions introduced in Table 2 above.

As we all know, for Frege there were only two ontological correlates of propositions: the True and the False. All true propositions denote the True, and all false prepositions denote the False. From an ontological point of view, if all true propositions denote exactly one and the same entity, then the underlying philosophical position is the absolute monism of facts.

In what follows I shall seek to disprove what Suszko called 'Frege's axiom': namely the assumption that there exist only two referents for propositions.

Frege's position on propositions was part of a more general view. Indeed, Frege adopted a *principle of homogeneity* (Perzanowski, 1993) according to which there are two fundamental categories of signs (*Bedeutungen* and truth-values) and two fundamental categories of senses (*Sinn* and *Gedanken*).

Both categories of signs (names and propositions) have sense and reference. The sense of a name is its *Sinn*, that way in which its referent is given, while the referent itself, the *Bedeutung*, is the object named by the name. As for propositions, their sense is the *Gedanke*, while their reference is their logical value.

Since the two semiotic triangles are entirely similar in structure, we need analyze only one of them: that relative to propositions.

Here *p* is a proposition, *s(p)* is the sense of *p*, and *r(p)* is the referent of *p*. The functional composition states that *s(p)* is the way in which *p* yields *r(p)*. The triangle has been drawn with the functions linking its vertexes explicitly shown. When the functions are composable, the triangle is said to commute, yielding $f(s(p)) = r(p)$, or $f \circ s(p) = r(p)$.

Fig. 1

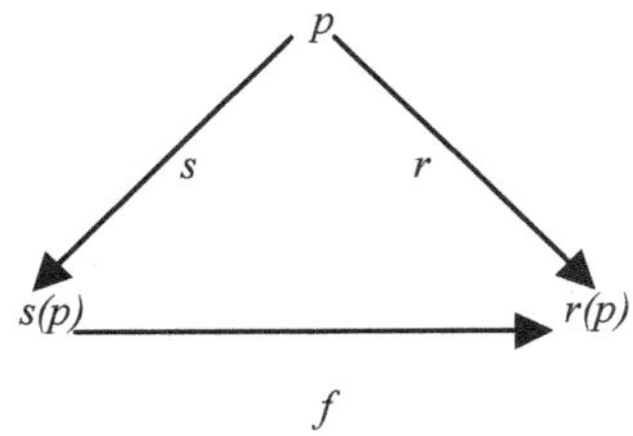

An interesting question now arises: is it possible to generalize the semiotic triangle? And if it is possible to do so, what is required?

Useful here are the Husserlian distinctions set out in Table 2. A first reorganization and generalization of the semiotic triangle therefore involves an explicit differentiation between the truth-value assigning function and the referent assigning function. We thus have the following double semiotic triangle:

Fig. 2

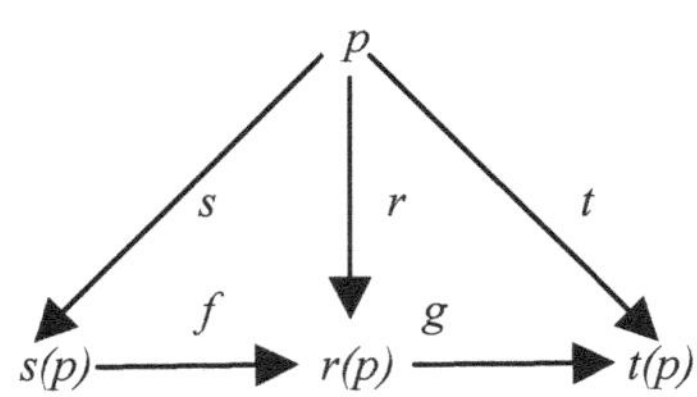

where *r* stands for the referent assigning function and *t* for the truth-value assigning function. Again on the basis of Table 2, we may further extend the original semiotic triangle by also considering utterances:

Fig. 3

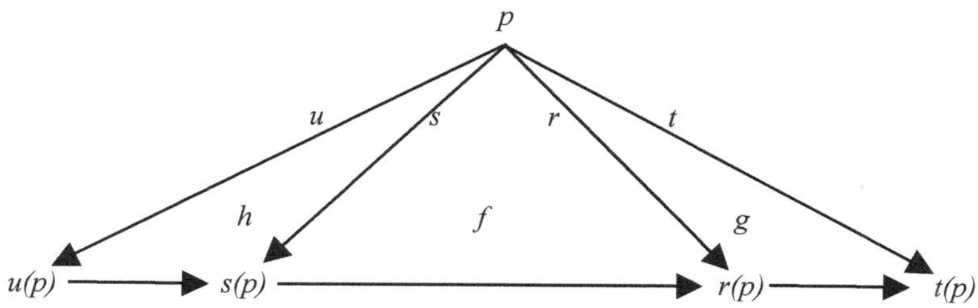

In order not to complicate the situation excessively, in what follows I shall deal solely with the right-hand side of the multi-triangle in Fig. 3, the one corresponding to what I have called the double semiotic triangle (Fig. 2).

Suszko uses the terms *logical valuations* for the procedures that assign truth-values, and *algebraic valuations* for those that assign referents (Suszko 1975). By arguing for the existence of only two referents, Frege ends up by collapsing logical and algebraic valuations together, thereby rendering them indistinguishable.

Having generalized the semiotic triangle into the double semiotic triangle, we must now address the following questions:

1. when do two propositions have the same truth value?
2. when do two propositions have the same referent?
3. when do two propositions have the same sense?

I shall distinguish among the three cases by talking respectively of *equivalence*, *identity* and *synonymy*. Sameness of logical value will be denoted by $\leftrightarrow$ (logical equivalence), while sameness of referent will be indicated with $\equiv$ and sameness of sense (synonymy) by $\approx$. Two propositions are synonymous when they have the same sense:

$$(p \approx q) = 1 \text{ iff } (s(p) = s(q)) = 1.$$

Two propositions are identical when they have the same referent:

$$(p \equiv q) = 1 \text{ iff } (r(p) = r(q)) = 1.$$

Two propositions are equivalent when they have the same truth value:

$(p \leftrightarrow q) = 1$ iff $(t(p) = t(q)) = 1$.

These various concepts are functionally connected as follows:

$s(p) = s(q)$ implies $r(p) = r(q)$,
$r(p) = r(q)$ implies $t(p) = t(q)$.

In general, the constraints that we impose on referents correspond to the *ontological assumptions* that characterize the theory. The most general logic of all is the one that imposes no restriction at all on *r* valuations. Just as Fregean logic recognizes only two referents so the most general logic recognizes a more than numerable set of them. Between these two extremes, of course, there are uncountable intermediate cases.

8 Conclusion

The items of the universe can be studied in various ways. The two that have proved most useful are *decomposition* and *parameterization*. The decomposition method presupposes that it is possible to inspect the item, and therefore that it is at least partly amenable to investigation. Ideally, the decomposition method works perfectly when the procedure is reversible and the whole can be reconstructed from its parts. The principle of truth-functional composition is a perfect formal codification of this ideal situation.

By contrast, the parameterization method treats the item as a black box which cannot be opened to see what is inside. But it is still possible to study the item by analyzing its responses to changes in external conditions (which in this case function as parameters). The most trivial example is that of parameterization with respect to time. In this case, the expression T → O can be used to indicate study of item O with respect to the parameter 'time'. The study of the relation of equivalence, identity and synonymy briefly outlined above belongs to this second method of analysis.

Bibliography

G.E.M. Anscombe, 'Under a Description', in G.E.M. Anscombe, *Metaphysics and the Philosophy of Mind*, Blackwell, Oxford 1981, pp. 208-219.

G.E.M. Anscombe and P. Geach, *Three Philosophers. Aristotle, Aquinas, Frege*, Cornell University Press, Oxford 1961.

N.B. Cocchiarella, 'Formal Ontology and the Foundations of Mathematics', in G. Nakhnikian (ed.), *Bertrand Russell's Philosophy*, Duckworth, London 1974, pp. 29-46.

N.B. Cocchiarella, 'Predication versus Membership in the Distinction Between Logic as Language and Logic as Calculus', *Synthèse*, 77, 1988, pp. 37-72.

K. Gödel, 'Russell's Mathematical Logic', in P. Schilpp, *The Philosophy of Bertrand Russell*, Open Court, Chicago 1944 (4th ed. 1971), pp. 123-153.

J. van Heijenoort, 'Logic as Language and Logic as Calculus', *Synthèse*, 17, 1967, pp. 324-340 (various reprints).

J. Perzanowski, 'What is Non-Fregean in the Semantics of Wittgenstein's *Tractatus* and Why?', *Axiomathes*, 4, 1993, pp. 357-372, previosly published as 'Ce qu'il y a de non Frégéen dans la sémantique du Tractatus de Wittgenstein and pourquoi', in J. Sebestik and A. Soulez (eds.), *Wittgenstein et la philosophie aujoud'hui*, Meridiens, Paris 1992, pp. 163-177.

R. Poli, 'Husserl's Conception of Formal Ontology', in *History and Philosophy of Logic*, 14, 1993, pp. 1-14.

R. Poli, 'Formal Aspects of Reduplication', *Logic and Logical Philosophy*, 2, 1994, pp. 87-102.

R. Poli, '*Qua*-Theories', in L. Albertazzi (ed.), *Shapes of Forms*, Kluwer, Dordrecht 1998, pp. 245-256.

R. Poli, *Alwis. Ontology for Knowledge Engineers*, PhD thesis, Utrecht 2001.

R. Poli, 'Ontological Methodology', *International Journal of Human-Computer Studies*, 56, 2002, pp. 639-664.

G.E. Rosado Haddock, 'On Husserl's Distinction Between State of Affairs *(Sachverhalt)* and Situation of Affairs *(Sachlage)*', in

T.M. Seebohm, D. Føllesdal and J.N. Mohanty (eds.), *Phenomenology and the Formal Sciences*, Kluwer Academic Publishers, Dordrecht 1991, pp. 35-46; republished in C.O. Hill and G.E. Rosado Haddock, *Husserl or Frege? Meaning, Objectivity, and Mathematics*, Open Court, Chicago and La Salle 2000, pp. 253-262.

K. Schuhmann and B. Smith, 'Questions: An Essay in Daubertian Phenomenology', *Philosophy and Phenomenological Research*, 47, 1987, pp. 353-84.

D.W. Smith 1991, 'On Situations and States of Affairs', in T.M. Seebohm, D. Føllesdal and J.N. Mohanty (eds.), *Phenomenology and the Formal Sciences*, Kluwer Academic Publishers, Dordrecht 1991, pp. 49-57.

R. Suszko, 'Abolition of the Fregean Axiom', in R. Parikh (ed.), *Logic Colloquium*, Springer Verlag, Berlin 1975, pp. 169-239.

Roberto Poli, Dipartimento di Sociologia e Ricerca Sociale, Università di Trento

Antonino Porrello

Theory of the Simulative Models (TMS)

Abstract:

Nowadays, for virtual reality, we intend the simulation of situations and emotions that is possible to be reproduced by making man interact with the computer in order to obtain the dematerialized reconstruction of real environments. This process revives the reflection on the importance of the technical-technological system. Because of its widespread diffusion, computerized information has eroded many expressive spaces. The astonishing development of consumer computer science, has made us speculate on the birth of a new model of economic development, based on the so called law of 'multiplicity', instead of the law of 'shortage': the digital revolution; the acceleration of the transmission system of information, knowledge and culture; the effects of digital revolution on the productive world are modifying the relationship between economy, work and society itself.

Keywords:
artificial, creativity, design, method, simulative model

1 Introduction

The images derived from synthesis, apart from their degree of virtuality – either weak or strong, window-like or immersive – they are mathematical models meant for a visual simulation of the objects and/or processes of the real world. They are abstract spaces capable of shaping intuitive or physical spaces. In the long story of the scientific modelling, the introduction of the virtual models of synthesis represents a true and actual turn. The traditional models were prevailingly visual analogies of a mechanical nature.

> Both the models of synthesis – virtual or not – and the traditional mechanical models have to reproduce the real, but in the first case, differently from the second, the image reproduced, resulting from the process, is not an arbitrary reproduction, or better, it is arbitrary only in a minimum part. This is explained by the fact that, while the traditional mechanical images are derived from a metaphorical choice, the images of synthesis are the result of a technical process – already present in the fields of photography, films, radiology – a process developed in direct generative contact with the object that is reproduced (Maldonado, 1997, 164).

At the beginning of the eighties the remarkable potential of modelling and simulation, provided by the computerized graphic art, opened new and unexpected perspectives in the field of medical radiology, both in the diagnostic component, in the therapeutic one, and even in the surgical one. This new evolution opened the way to sensational technical-scientific developments, that having recourse to the technologies of ionizing or not-ionizing radiations, contributed in a richer and richer knowledge of a universe that a somatic opacity had always hidden, yielding some of its secrets through invasive actions. There was still an unresolved problem: how to translate this knowledge into three-dimensional models or simulations to intervene operatively, or rather interactively and in real time, on the obtained images. Nowadays medical imaging can rely on visualization tools and technologies for the modelling of solids; it also permits the observation *in vitro* and intervening, by means of a dynamic-interactive monitoring in a three-dimensional space, on the structures and functions of the human body. Tomàs Maldonado observes as the strong innovative impact of the virtual-interactive modelling made itself known, during the last years, not only in the medical disciplines (or specializations), but also among companies that are getting ready to make use of such models in the process of products industrialization and to propose an interactive-direct connection with the final consumers of the products themselves. The company, as in its nature, has been making questions that, up to now, did not find unanimous reply between the scholars concerned with the theoretical and practical aspects regarding the images generated by the computer. If the point of arrival is a digitalized image, the point of departure is the extraction of an image from a real object. It is understandable that between the virtual image of synthesis and the image extracted from the real object there is a high degree of similarity. A degree of similarity that, in the last times, thanks to the astounding technical progresses in the field of the virtual modelling, has highly increased. Though, as the story of the scientific modelling teaches, it is difficult, if not impossible, to talk about similarity of a model with respect to the reality without having to confront, on the theoretical level, with the immense domain of questions that the idea of similarity has always raised. And the reason is simple: in the past no models of scientific visualization claimed to act as the twin of the real model.

By the end of the nineties some projects of industrial research have turned to the study of technologies for the virtualization of anthropometrical integrated models through three-dimensional models in motion and models for the simulation of the industrialization processes of garments. The introduction of such technologies is to be considered as a tool meant to revolutionize not only the industrialization of the product, but also its strategy of commercialisation and the consumer's satisfaction themselves.

The study of the concepts and the methods through which we get an understanding of the scientific imagination and the connected technologies, employed in the operative research in the technology of simulation and 'physically-based' and 'creation images' technologies, turns out to be of a strategic importance.

The use of the virtual technology presupposes the formulation of models and representations of reality that aim at being reproduced in the virtual reality and, as a consequence, it is necessary to explain what are precisely the limits, the expectations and the state of development of such models. The high potentiality represented by the modelling and the simulation of the phenomena, stands essentially, on one side in the capacity of devising a proper methodology of research to test the knowledge of physical phenomena; on the other in the possibility of escaping the two classical dimensions, that is space and time. When we start monitoring the simulation, this opportunity lets us have a simulation in the real scale, the one we are used to, that is the scale of our physical experience, and in a larger or smaller scale. This procedure has also a set of implications on our conception of time, on the possibility of varying its direction. We live within a scansion of the time determined by our physical life; entering a virtual world of simulation we have the possibility of accelerating or slowing down the phenomena. Therefore we can see and understand phenomena of synthesis that have a very long duration or understand individual steps of phenomena that have an extremely short duration. This process increases our capacity of investigation.

2 The approach to the simulation of situations in order to get a dematerialised reconstruction of the real processes

2.1 A. The Theory of the Artificial (TA)

The Artificial, in its version introduced by Massimo Negrotti (2000, 1999, 6, 244), is made real in a process that aims at reproducing a 'natural process', but also in a project or in the construction of a device that 'reproduces some natural objects'. An object or a process are defined artificial if they have been built by man (necessary condition) with the purpose of reproducing either an object or a natural process (sufficient condition) by means of materials or different procedures (prerequisite by definition). Negrotti defines the Theory of the Artificial in terms of verifiability of the scientific theory, and this presupposes: a) that there is a theory (scientific activity as knowledge of the world); b) that science is capable of creating objective situations (this is the task of technology) so that the experiment (or the empirical process) can be used.

The problem, which Negrotti refers to, is born in this process of methodological definition. This is evident in Negrotti's statement (1999, 244): 'The TA (Theory of the Artificial) proceeds by abduction and describes some features of observation on the base of the results of any *artificialistic* process'. This statement creates the condition for us to have to find a beginning to start out and formulate a scientific hypothesis. If by abduction we mean the process of objective observation, from which a sort of theoretical estrangement (abduction) springs out (the one anticipating the formulation of the hypothesis), then the reference is to the principle of induction, that is observing a series of phenomena, all of which supplying a unique indication, and from these data (through a logical leap) elaborating an hypothesis. Therefore: 'Such results are characterized by a selection of performances, if compared with the natural specimen' (1999, 244). The artificial appearance therefore is nothing more than the creation in a laboratory (by the use of technological tools) of real and optimal situations, in the meaning of selected and controlled, for the experiment, to test the validity of a scientific assumption, that is to promote a new knowledge of the world.

Negrotti states that the artificial is 'the result of a deliberate [...] multiple selection' (1999, 244). 'The Theory the Artificial defines this selective feature of reality, as *the choice or the construction of levels of observation* (LO)' (1999, 245 italic in the text). The level of observation, in this case, can be explained as the methodological process that searches for the greater number of cases (or simulated situations) capable of supplying a wider set of knowledge. An example as old as the Aristotelian philosophy is the following one: if I see only one white swan, I think that all swans are white; if I see a hundred white swans I give it for certain that swans cannot but be white. The last clause can be considered as a sort of Level of Observation. The philosopher however in the end says that he saw one and only one black swan: does the identity between white and black still exist? Obviously not! By this example I would like to explain Negrotti's consideration that observation is characterized by a process of observative selection. The Theory of the Artificial defines this selective feature of reality, as the 'construction of levels of observation' (1999, 245). A lot of centuries have passed since then, nowadays the swans are the laboratory experiments, that is to say the artificial realities by means of which we proceeded to observation; however the logic does not change. As a consequence this reasoning leads us to the consideration that 'The LO (Level of Observation) is a procedure or a profile through which the researcher describes the process that he is trying to generate in an artificial version' (1999, 245), similarly to the assertion from which the researcher is willing to start. At this point a reflection is born, or better the choice of a methodological system of testing of the theories originated from the different levels of observation. At this level the system of choice intervenes following a series of experiments that corroborate a theory or other experiments that justify another one.

Massimo Negrotti's essay, through a long reflection, stresses the importance of the technical-technological system[1], of the real possibility of realizing an objective situation (therefore a real situation)[2] and the assertion that 'the representation of the reality [...] soon turns out to be the true exemplary of the artificialist, [...] a generation of representations as a result of the observation of the world [...]' (1999, 248). Eventually, to further justify the last consideration, there is Negrotti's assertion that perception (meant as the capacity of knowing in an epistemological sense) becomes observation no longer of the actual reality, rather of the laboratory reality, that is the artificial one that we were referring to here above. The difference between observation and perception is really singular since it distinguishes on one side what is proper of senses, perception, from observation, the logical process that guides every real scientific knowledge. We can affirm, with a definite degree of certainty, that every scientific demonstration presupposes a theoretical knowledge (though not verifiable) of the natural phenomena[3] and that every demonstration (and this is the capacity of creating an artificial reality, nowadays called virtual reality) can be only accepted in the logic phase. This is one of the complex problems that technology can help to solve, but at the same time it creates some complications. The true meaning of technology, as stated by Antonino Zichichi[4], is the application of science: the progress of science intruding into the technological progress of development. Its application to practical business aims, in its version of virtual simulation, has been made possible thanks to the convergence of various technologies

1 I am still convinced that technology actually is science and that everything that is not theoretical is only technical. I suspect that only a few agree with me! And if I asked those who disagree, what is technology, the study of technique and its application? Probably I would receive an evasive answer, not too different from the definition of science, that is the set of results deriving from the speculative human activity meant for the knowledge of causes, laws, and effects concerning a specific order of phenomena and based on the method, the study and the experience, that is experimentation through technology.

2 I believe that the core of the research and book is centred on this wide and interesting theme.

3 Bohr's atom is nothing but a small example.

4 Lecture given during the Meeting 'Virtualità e Modellistica', Treviso, 1999.

among which: mathematical models, models programming, databases, data acquisition sensors, integration and manipulation of the sensors and the connected information, and communications. Indeed, the possibility of extending the boundary of modelling to the market of services and entrepreneurial products depends on the possibility of permitting an easy usability of models and complex mathematical systems.

2.2 B. Creative process

Bloor (1976) and Mulkay's (1981) studies removed from scientific knowledge the aura of sociologic 'extraneousness', making the event, which is scientific in itself, a sociological relevant phenomenon, observable and therefore the object of a sociological analysis.

By recognizing the product of knowledge as an act of creativity, we attribute a substantial importance to the 'social context of reference: condition of the creative process'. The acknowledgement, indeed, qualifies the subjective creative experience as an 'original contribution to the common heritage of knowledge' (Neresini 1994, 64).

There are three levels of observation linked with the understanding of the creative process of the same phenomenon:

1. the scientist in his direct activity of research;
2. 2.1 the scientist in his reflective position towards his own actual activity of research;
   2.2 the subjects (such as the sociologist) which are part of the process of research, without directly taking part into it;
3. the subject using the created product, in the case of applied sciences.

The distinction between the creative experience characterizing the scientist in action, and that affecting the attribution of creative value to the products deriving from such experience, allows us to explain better the matter of the 'dependence of scientific discovery from the social context of reference' (Neresini, 65). The difference of meaning undertaken by the creative process between the scientist that enacts the process and the context that judges the values of his results, is well known.

The creative process (the new connection that generates the intuition) develops itself through a series of key passages (Neresini, 1994, 66)

1. the identification of the problem to be solved, deriving it:
   a. directly from the literature available in a certain area of research;
   b. by transfer from a different area of research;
   proposing it in new configuration. This second case can move from the definition of the problem that we are willing to examine. 'The definition of the problem itself covers in itself a large part of the creative path', given that the solution of the problem also derives from putting the question in different terms;
2. once we define the problem, we can treat it having 'recourse to procedures and strategies of known solutions'. This phase can give place to repeated negative results that look like the accumulation of elements of knowledge, contributing in the rising of intuition;
3. having recourse to an alternative method, consequent to a phase of pause; to a different valorization of certain known elements provoked by more or less casual circumstances, not least the moment of estrangement from the direct activity of research. At the same time the chance in the occurrence of the event is strictly connected with its continuity and dependence from the socially available knowledge on the matter, provided by the scientific community.

The results produced by the group of research require the recognition of the context of reference – the scientific community or the institution, the action in the applied researches – because the creative process can be considered concluded[5].

2.3 C. Methods of simulation

The objective of this study is an analysis of the methods that can be used to set reality free of its materiality to build it and to rebuild it on the basis of the operator-user's interest; the degrees of technological

5 A hypothesis of working, emerged during the analysis of the KAEMaRT case study, was concerned with creativeness in terms of quantity: degrees or percentages interested by the change.

innovation introduced to get the virtual representation of reality taken into consideration.

The technologies of modelling and simulation of non-rigid materials were used for the computerized animation, applied to the industrial areas whose production is characterized by objects of design and flexible functions, as for example the clothing industry. There are different methods to model and simulate the behaviour of non-rigid materials: the method based on the particles (particles model) and that based on the finite element. The first one has been used by the graphic computer, for animation, a field in which visual realism is very important.

Both methods, though with different purposes, are applied to the textile and clothing industry. In these fields a special attention is paid to the physical accuracy in the study of the behaviour of materials, in connection to their mechanical properties. A careful activity of design was carried out to produce some virtual representations – such as material modelling, technologies to speed up the calculation of the time, the search for collisions, algorithms, etc. – daily use's tools. To reach this aim, the evolution of simulation, through the introduction of new models of interaction and new technologies in the phase of simulation, is oriented towards the use of operative procedures of high level and more user-oriented.

The theoretical aspects are twofold: one is concerned with innovation in a wide sense, that is to say in the technology and theory (where theoretical aspects exist) that support innovation with respect to the existing state of the art. The second innovative aspect manifests itself when we develop or modify an existing technology. The transfer of technology from an area to another one, brings about innovation in an absolute and a relative sense: in a relative sense since it applies to the sector under examination; in an absolute sense because the transfer involves the development of the existing technology at different levels.

In the last decade of the XX century the scientific and technological research together with industry, with renewed vitality, have oriented towards the modelling and the simulation of deformable objects. There are two technologies used for the modelling and simulation: the technology based on geometry and the hybrid technology based on the material and the physical aspects.

The most proper modelling technology to describe the non-rigid objects is the physical and material model, commonly used. In fact 'the physical model is a mathematical representation of an object and its behaviour that incorporates forces, twisting, energy and other attributes of the Newtonian physics'. It is a model that 'reacts in a natural manner to the external forces applied, such as gravity, constraints and the impenetrable objects', in the same way as we expect a natural object to react. (Eurographics, 99, 2)

The physical model, in turn, can rely on two models: the continuous and the discrete ones. The discrete model, which, as we will see later has been used by KAEMaRT, 'describes a deformable object combining very simple mechanical elements, in which all types of logical and non-logical interaction are managed in the same way and the simulation is carried out by calculating the laws of interaction between the objects. In this manner the object is discretized in small elements' sharing a kind of connection between them. (Eurographics, 99, 5) The discrete model lets us describe an object by means of a set of particles provided with physical properties (mass, ray, and other properties).

Another technology is the energy-based technology. 'This technology calculates the energy affecting the whole system starting from a set of equations, it determines its shape moving the points so as to reach the smaller state of energy' (Eurographics, 99, 6)[6]. The model lets us face the problem in its entirety and it simplifies its optimization: simplicity, generality, possibility of specifications of the global properties; management of interactions with the external world in a quite uniform manner.

6 An interesting work has been conducted by House and Breen, where they observe the mechanical system, constituted by the fabric, in which the energy concerning the folds and the cut is calculated from data obtained using the Kawataba system. Quoted in U. Cugini, M. Bordegoni, C. Rizzi, F. De Angelis, P. Prati, *Modelling and Haptic Interaction with non rigid Materials*, Eurographics, 99.

3 Social Construction of the Scientific Creativity

The theory of the technological discovery is based on the hypothesis that: 'there exist principles of invention which constitute the basis of technological invention'; 'logical processes that carry to inventions'. The aspects (problems) of designing can be ascribed to two concepts: knowledge and *'cultural indolence'*. Knowledge is linked to the innovative principles that characterize the solution of design problems; to ideas and concepts that are already present in other projects, used in an alternative way. 'Psychological indolence' characterizes the researcher's unavailability to take in consideration the exploration of different alternatives characterizing the universal principles of invention. The method that we call '*Theory of the Simulative Models* (TMS)', can be a tool to know the creative technique, structured in two concepts: a) knowledge as the technical exploration of ideas and concepts that have been already applied to other solutions; b) design meant for the identification of innovative solutions. The method is based on inventions, on the analysis of the results of creativity in the technology. These results are derived from the technologists' activity, by the real participation in the technological research that is based on observation and calculation, on the rational necessity of connecting each detail to the whole design. The exam of the models of the scientific imagination inevitably appears characterized by an inductive and an empirical quality, entrusted to a scrupulous participation in the process of technological innovation, to a historical accuracy and to a careful study of the documents and the available testimonies.

The study of the scientific work asks for a methodology of research to be shared with the recourse to a 'species' of ethologic approach, as proposed by Gerald Holton (1973), and before him by Peter Medawar (1967), a scientist interested in the problems of knowledge, with the help of the technology of inquiry, proper of thematic analysis. The *thematic analysis* (similarly to the content analysis) lets us emphasize some constant and continuous aspects in the development of science and the same scientific work produced by the person or group of people whose work is being studied.

4 Theoretical Formulation of the Simulative Models (TMS)

Substantial object: *what do we want to realize?*

4.1 I – Simulation aims

1. Theoretical aims: Theory of the scientific imagination (H. Golden, 1973), Theory of the Artificial (TA) (M. Negrotti, 2000);
2. Methodological aims: regarding the feasibility of a project of simulation;

How the system works:

3. Anticipatory aims: what happens applying to the system specific inputs, that is manipulating the starting variables;
4. Innovative aims: with regard to the solution of a practical problem;

How we want to realize:

1. Innovative technology *tout cours*;
2. Innovative technology for an external transfer – from an area (productive or software) to another with the elaboration of original forms;
3. Innovative technology for a transfer within the same area with further original elaborations;
4. Innovative technology in relation to the use of technologies or software with adaptive interventions;
5. Innovative technologies in relation to the use of elementary software of an applicative nature, through with original aims.

4.1.1 Levels of invention:

- Level I: the solution adopted does not alter the system;
- Level II: the system has been modified in some ways;
- Level III: the changes introduced in the system, or in a sub-system, or in a significant part of it, are radical;
- Levels IV and V: the introduced changes are discoveries based on previously unknown information.

In the case of the inventions of:

- Level I: the solutions come from the designer's field of speciality;
- Level II: the solutions come from the same branch of technology;

- Level III: the solutions come from the technologies linked to the problem;
- Level IV: the solutions come from the fields of physics, chemistry, and geometry – usually from effects and little known phenomena;
- Level V: the solutions come from beyond the limits of the contemporary science.

4.2 II – The Functioning

The technology supplied by the mathematical algorithms for the representation of the connections: The Computer Technology[7] (G. Natalichi, 1994; R. Counts, in L. Ricolfi, 1997, 285-315; Cecchini, Indovina, 1989)

4.3 III – The Acquisition of Information

1. Procedure and tools through which the new models of the collection are realized: What information are necessary to realize what we want to obtain.
2. Building up and realization of a new cloth by means of certain changes to the finished product: in applied research applied the question is directed to researchers and designers on one side and to the Customer/user on the other.

4.4 IV – The Realization of the Simulative Model

1. virtual design of the reproducible universe;
2. optimisation of the objectives;
3. construction of the project: functions and esteem of the parameters;
4. solutions of the problems of the project (A. Bruschi, 1999).

7 R. Conte, in L. Ricolfi, *La ricerca quantitativa*, Angeli, Milano, 1997, p. 285-315.

5 Construction and use of the simulative models

1. Construction of the *Levels of observation* (Negrotti, 2000, 1999), of the reality that we want to simulate;
2. Translation in the *Model of the physical phenomenon*;
3. Formulation of the *IT Model* capable of simulating the physical model;
4. Execution of the consequent operations;
5. Attribution of the obtained results to the reproducible universe through simulation;
6. Manner of use of the realized product;
7. Manner of evaluation of the proposed product (validity of the obtained results);
8. Possible changes brought about in the project.

6 Simulation Technologies – The Case Study[8]

The KAEMaRT Group is formed by senior researchers with multidisciplinary skills: a Mechanical Engineer (Prof. Umberto Cugini) – the Group coordinator – a Physicist (Prof. Caterina Rizzi) and a Computer Engineer (Prof. Monica Bordegoni), besides a staff of junior researchers and technicians who collaborate according to special projects or needs. The researches conducted by the KAEMaRT Group have been directed towards the virtualization of the management of the sales-samples, taking in consideration two factors: 1. the choice of the garments that will constitute the sales-samples is a consequence of a considerable proposal of items; 2. the considerable long time that runs between the moment in which the construction of the sales-samples begins and the moment in which the garments are ready for the customer. In case the production line is particularly lengthy, this interval could last more than one year. As a consequence the company requires a reduction of the physical prototypes to be realized and a reduction of the time necessary for the product to reach the market.

8 For the presentation of the case study, apart from the study of the documents written by the KAEMaRT Group that Umberto Cugini let me freely access, I also relied on a deep interview to Prof. Umberto Cugini and Caterina Rizzi.

The object of the projects of research in the clothing area has been the 'development of a virtual environment for the 3D design of garments', with the purpose of:

- defining three-dimensional models of the garments;
- carrying out the automatic laying-down;
- simulating its behaviour, taking into account the mechanical properties of the material/fabric employed in the production.

The researches were directed towards the development of 'a graphic and interactive tool, that could allow the modellist/designer to design male and female items of clothing in a 3D environment and *to monitor and reproduce, on the base of the parameters that characterize the physical model, the real behaviour and, as a consequence, to reduce the number and the role of the sample/physical prototype*'[9]. The targeting of the results of the activity of research is the reduction of the times of execution of certain phases of the process of industrialization.

The KAEMaRT Group developed a system called 'SoftWorld' particularly directed to the industrial applications[10]. Taking in consideration the fact that in the industrial areas it is not sufficient to produce simulations that seem real, but both the physical accuracy and the visual realism are also necessary, SoftWorld adopts a particles model. According to this model an object is discretized in a set of particles connected by forces that characterize the mechanical behaviour of the material. The main component of the system is the simulator which includes a numerical ODE solver, based on the Eulero method; and a manager of constraints and collisions[11].

9 'Ricerca Industriale nel Settore Abbigliamento: Progetti e Risultati', lecture presented during the Meeting 'Virtualità e Modellistica', Treviso, Novembre 1999, p. 1.

10 U. Cugini, M. Bordegoni, C. Rizzi, F. De Angelis, P. Prati, *Modelling and Haptic Interaction with non rigid Materials*, p. 9, sez. 3.2. Constraint Management, Eurographics, 99; A. Frugoli, R. Rizzi, 'Physically-based simulation with SftWorld', in *Proceeding IASTED International Conference on Applied Modelling and Simulation*, Honolulu, August, 1998, pp. 373-377.

11 Cugini and others, op. cit. p. 7.

For the management of constraints it has been used the method of Dynamic Constraints[12] since it permits the application of multiple constraints to the same particle and to ensure the respect of every constraint at each step, in each phase of the simulation. The management of collisions is based on the technology of the *Bounding box hierarchy technique*[13].

The cases that were tested, asked for the simulation of different types of non-rigid bodies, ranging from fabrics for the textile and clothing industry, to wires or products for the food sector. The system has been partially integrated with a cinematic simulator[14] 'in order to obtain an integrated system to simulate the processes of production that involve non-rigid products' (Eurographics, 99, 8, 9).

The KAEMaRT Group participated, at first, in the European project MASCOT (funded by the European Community), concerning the woman's area. Among others, three end-users participated in the project. They had to evaluate the modules and the environments developed by the other partners (GFT Donna, Confeciones Mayoral, a children garments business in Spain and DCTA, a business in Great Britain)[15].

The second project, within the Consorzio TA2000, funded by the MURST's (the Italian Ministry of Universities and Scientific Research) National Textile Program, coordinated by the business F.lli Corneliani Spa, had as its object the simulation of a garment.

During the Mascot project the designing Group started the technological development of the women's clothing. Taking in consideration the traditional procedures for the realization of prototypes, they defined two lines of actions of research. One of this lines, the most innovative, foresaw the development of a 3D modeller that allowed the de-

12 Ibidem, Collision Management, p. 10.

13 Ibidem, Bounding box hierarchy, p. 11.

14 Cugini and others, cit. Eurographics 99, p. 8, 9.

15 F. Carignan, J. Yang, N. Magnenat Thalmann, D. Thalmann, 'Dressing animated synthetic actors with complex deformable clothes', in *Computer Graphics (Proceeding Siggraph)*, vol. 26, n. 2, July 1992, pp. 99-104; G. Volino, N. Magnenat Thalmann, J. Jianhua, D. Thalmann, 'An evolving system for simulating clothes on virtual actors', in *Computer Graphics in Textiles and apparel, IEEE Computer Graphics and Applications*, vol. 16, n. 5, September, pp. 42-51.

signer to operate directly on the garment, in a three-dimensional environment. Once they automatically obtained the 2D clothes and defined the assembling rules, the simulator generated the physical model and therefore it could simulate its behaviour.

Through the second line of research, the application to the men's sector, (Consorzio TA2000), the designers continued the development of the technology of simulation that, maintaining the business's historical background (in the specific case this means that they maintained the 2D systems), it foresaw techniques that, by using 2D clothes, assembled them on the dummy and afterwards passed the information to the simulator for the evaluation of the garment drop and the dummy's behaviour itself. This second application developed both the modelling module of the garment, and the simulation module. Compared with the woman's project, the new system was further modified and some new functionalities, concerned with the simulation of the men's garments, were added.

7 The Simulation Model

The simulation model used for both the projects was essentially concerned with a modelling environment, used to define three-dimensional models, 3D garments and the corresponding 2D clothes; a simulator; a database of dummies and materials (to correctly characterize the material and to confront and simulate the same garment using different types of fabric). The technological research has been oriented to the model of simulation, since it represents the crucial environment to reproduce the dynamic behaviour of the body. The functionalities distinguishing a system of simulation, in this context, can be synthesized as follows:

1. the creation of the physical model, in the specific case, by using a particles model (similar to the one used by prof. Thalman's group in the creation of images)[16] also including the description of the physical properties of the material;

16 In the lecture 'Ricerca Industriale nel Settore Abbigliamento: Progetti e Risultati', presented during the Meeting 'Virtualità e Modellistica', Treviso, Novembre

2. the assembling of the 2D clothes on the dummy, the use of the data generated by the 3D modeller, the clothes sewing, taking in consideration the seaming lines, their positions, the presence of darts, the execution and the simulation.

To grant the accuracy of results, from the physical point of view, a rectangular discretization has been used, oriented according to the fabric weft and warp.

The research project was preliminarily concerned with the process running from the ideation to the realization of the garment; after that the Group was involved in the creation of the garment: the choice of a style is an operation carried out by the stylist, starting from an existing garment, to get a new one, or an operation that can be carried out by a modellist to create a new garment.

The object of the research has been the identification of the possibility of realizing a virtual prototype, during the phase of the production of the sales-samples, therefore still far from the industrialization process.

During the first phase of development of the research, the way through which the company uses the garment have been analysed. A particular attention was paid to the process of tailoring creation which represents the real environment that the simulation process intends to reproduce. The precise operations necessary to the creation of a garment have been analysed in detail. The whole activity has been called 'definition of the system requirements'.

Once the explanation on the methodology of intervention were completed, the Group started the observations on the creation of a new garment and afterwards, the single operations describing this creation were analysed. The classic-cut jacket was chosen.

1999, Caterina Rizzi distinguishes between Thalmann's situation of visualization [Thalmann is the director of Geneva's MIRALAB and the Swiss Federal Institute of Technology's Computer Graphics Lab] and the situation of design, affecting KAEMaRT Group's experience. In both contexts the particles model has been used, but the crucial aspect is its employment. In animation projects the particles can be freely disposed according to the designer's will, whereas in the Group's design, they have been placed on a rectangular grid, because in this way they could obtain a characterization along the two directions.

For its realization the company chose a garment, whose model had been already developed and that could be undergo modifications such as lifting the shoulder line or tightening the waist or lengthening it, etc.

The observations were concentrated on the manner through which the modellist manually realizes the modifications. This first phase in the process was used by the KAEMaRT Group to understand which were the procedures adopted by the company for the realization of the garments to be included in the sales-samples and how this process belonged in the productive process.

The realization of a F.lli Corneliani collection taught the Group on the designing aspects, their representation in the show room, and the choice of the garments to compose the collection to be afterwards reproduced and sold. The higher attention was paid to the analysis of the composition of the collection and the development of the prototypes, since the Group was interested in understanding how the company developed a new garment.

When a new collection must be sketched or when some prototypes must be realized, the company relies on a procedure. A market survey is conducted to define the structure of the collection and the draft of a technical card concerning it.

For the realization of the prototypes the procedure gives the following instruction: to choose the starting model, that is the garment of reference, to make modifications on the paper pattern and in the end to reproduce physically the prototype.

The designers were interested in the determination of procedures and tools used by the company to produce a new garment. To this aim the structural elements of a jacket were determined, that is its main constituent parts, and the modifications to be brought about. Once all these aspects had been considered, it was crucial to determine how these analytical results could be correlated in a three-dimensional sphere of action. The first phase of the designing process was concerned with the observation of the procedures followed by the company, in order to prefigure how, through the new process of simulation, the stylists, the tailors and the modellists, could work. The second phase was concerned with the analysis of the procedure and the single activities carried out in laboratory to modify the tailoring garment of reference to get a new one.

The procedure adopted by the designers can be described as follows: to observe the tools for the realization of a new garment without analysing in detail the single operations; to observe the manner of working at time '0' to determine how the group would enter their process and how this would change; to evaluate which functionalities would have satisfied the simulation system and how to render their operability from a three-dimensional point of view. The further widening was concerned with the modifications to be realized on the garment to get a new one: 1) the company started from an existing material prototypes and modified it. On the basis of the modifications brought about, the technicians developed a new paper pattern and realized the prototype; 2) evaluation of the single modifications, that is, definition of the requirements.

These two phases have been designed to understand which were the functionalities that should be possessed by the system.

8 What are the components constituting the system (essentially by the functional point of view) and how are they used by the designer?

The description of modules (the components of a system) lets us observe how the system will be used by the designer. Indeed the system is composed by a series of components that will permit the use through operations.

During the design phase, the group realized well-determined simulations, under the conviction that the project of simulation represented a part of the real industrial process, in which it should fit. When the group was concerned with the 2D clothes, they realized that, though initially they thought of realizing them automatically out of the 3D model, they had not taken in consideration the company's historical background. It was then necessary to assure consistency between the results of the designing phase and the 2D models to be realized to industrialize the garment. Therefore they had to change the formulation of the question: the automatic laying-down of the 2D clothes was substituted by the use of information deriving from the modifications carried out by the company system. In this manner the operators would automatically provide for the laying-down, using their company's commonly used procedure. Given that the company realized the garment making modifications on an item of clothing of reference, the KAEMaRT Group deemed it necessary, in order to simulate this procedure, to create a three-dimensional representation of the garment, that is a geometrical model of the item of reference, a representation that could be carried out by the Group.

9 Innovative Design

The designing phase has been carried out, by using technologies such as the reverse engineering and the acquisition of data by physical models. A physical model has been digitalized, in a way to obtain a set of points (called a 'cloud of points') out of which a geometrical model was generated, to be imported, afterwards, in the Maya system. Thanks to some of the potentialities of the Maya system, it was possible to create a module provided with an interface to make modifications on the garment according to the procedures applied by the company. The Group employed the reverse engineering technology to this case, a technology already used in the motor and aeronautic sectors. The transfer required the solution of problems emerging during the acquisition of data by the physical model. This solution brought about some modification to the technology in order to get the position of the space and the garment (jacket) and all its points, and also the disarticulated positioning of the physical garment, taking in consideration the digitalisation system in use. The present technology, actually, operates on rigid objects, whereas the Group had to deal with soft objects (fabrics) and also further information, for instance on seams, had to be found.

In the realization of the Project Men's Clothing, the KARMaRT Group, once they acquired the data by the company, developed the modeller. The development of the 3D modeller required a two-phase activity.

The first phase was the acquisition of data concerning the jacket (the 3D model of the jacket and the dummy's model). To reach this aim, two systems have been used: Athos, a system for the digitalisation of the jacket and the dummy, and Paraform, a system to generate the geometrical model, starting from the cloud of points.

The second phase was concerned with the development of the 3D modeller and in particular the Module 'Modifications'. This Module is used to make modifications in the 3D geometrical model of the jacket. The modeller elaborated by KAEMaRT is an interface allowing the designer to work according to his own habit. Once the modifications are made in the base-garment, they are transferred to the company in order to derive the 2D clothes which form the jacket, within the company system.

> The modeller required the introduction of two interventions: the acquisition of data (digitalisation and revision of data) and the development of the software, by the use of the available technologies, properly modified. The software developed by the KAEMaRT Group employs mathematical models. Different technologies were used to acquire 3D data and to develop the modelling software. For the development of the modeller, and in particular of the Module Modifications and the interface-user, they decided to acquire the commercial software MAYA produced by Alias/Wavefront, that was the base for the development of the system.
>
> The development of the module for the 3D simulation of the designed garment entirely derived from a technology exclusively elaborated by the KAEMaRT Group. They already disposed of a simulator for non-rigid, deformable objects such as fabrics and they adapted it for applications to men's clothing. The biggest problem was the transfer of data generated by the company through its CAD 2D system to carry out the simulation in a correct way. They adapted some parts of the 3D simulation module in order to answer this problem. The software was a tool to simulate non-rigid object and was based on a particles model. The problem was the passage from the 3D modeller, by means of which the object was represented through triangular parts, to the representation through particles. It was necessary to transfer information, and this was done by planning new functionalities such as the automatic creation of the physical model through particles of men's clothing.
>
> The Group created a software capable of converting data of different formats in order to match the company requirements. Afterwards they created the 2D models and they were positioned on the dummy. This operation of preparation

of the model to be simulated was exclusively designed for the men's sector. The starting point was the simulation model for women's clothing that they already disposed of, and they adapted it to specifically suit the men's sector. This software-base capable of simulating the objects behaviour (the core of the whole system) was enriched and integrated to specifically suit the men's sector of clothing.

To carry out the simulation it was necessary that the object was represented by a particles model. Therefore to generate a particles model of the garment it was necessary to develop ad hoc modules for the generation of the physical model through particles starting from the 2D clothes generated for the item manufacturing. The simulator core had not been modified.

The Group always based his work on techniques, methods, algorithms already available in literature. They somewhat improved these techniques, used in a unconditioned way above all in the animation and cinematographic sectors, because they would respect precise physical constraints, that, in the case under examination were determinant. The aforesaid technicalities, in some cases, had also been used to deal with fabrics, nonetheless they were oriented towards problems of animation. The more involving designing activity was done for the generation of the real physical model and the positioning-assembling of the garment itself. These activities were necessary for the simulation of the garment behaviour. On the contrary, the simulation activities were quite unchanged, they only were optimised.

Once the modeller was ready and the studies on the simulator started, the necessity of integrating the two system emerged, so that the designer could use them together. Therefore a unique integrated environment was set up enabling the modelling, the generation of data, the connection with the company system in order to generate 2D clothes and, eventually, the activation of the simulation, once the 2D clothes were realized. The modeller now being ready, the KAEMaRT Group gave it to the company, so that it could evaluate it, while the Group carried on the simulation experimentation.

During the first phase the simulator worked consistently with the data provided by the company, but the adopted real physical model resulted too simplified, even though they had taken in consideration the presence of some structural parts composed by different layers of material. Problems arouse during the simulation of the garment and particularly when some undesired effects were identified in the neckline or in the lower part, that is near the vents, etc.

For the generation of the physical model of the garment they had started from the same premises hypothesized for the women's sector, according to which, even though in some parts there was the lining or at the waist level there should be a higher stiffness, the fabric should be treated as a single sheet, and optimal results would derive. The only thing that had been differentiated for the women too, was the part corresponding to the breast and to the shoulder

straps, that were parameterized in a different way. The particles were the same, but the links were characterized in a different manner, because the stiffness corresponding to those parts of the body were different. Therefore a new starting physical model was created where the garment was divided into numerous sub-parts that did not only correspond to the 2D clothes.
For example, the 2D cloth corresponding to the breast had been divided in an upper part, a central part, a bottom part, a side part (where the lining and the hem are positioned, etc.). All this helped the group understand that a jacket is also made up of linings, hem, buttons, etc. The second phase was characterized by changes: the modification of the software and the one concerning the starting physical model through different parameters.
At this point the simulation had improved, however the problem of the lack of experimental data, that is the measurements concerning the fabric features with the lining the hem, etc., kept unresolved. The project had almost reached its end and the only part developed was the front of the jacket (and its various layers), though the identified sub-area were still missing.
To test the validity of the approach, that is, if really the one they were following was the right way to be followed, they deemed it proper to develop a new environment basing on the simulator which permitted the modification of the single areas. This program could permit the user to make evaluations and determinations such as: if the neckline is too soft, it is possible to stiffen it, to remake the simulation and to evaluate the effect. In this manner they could understand if by this approach, according to which they divided the jacket into areas, rather then in components, they could get better results in the simulation. The system to modify the area of the jacket, called 'Textile Laboratory', though in an empirical phase, was the result of a collaboration between the designers, that is the KAEMaRT Group, the company's stylists and modellists. It is still an experimental part that must undergo steps of verification with the system 'Kawabata' to get real data and to evaluate if the obtained data correspond to the real ones. This system requires new data to automatically elaborate the new model (without re-defining it from the beginning) and the new simulation. Also the 'Textile Laboratory' is the result of a new technology elaborated by the KAEMaRT Group using as an interface elaborated by the software called Java. The 'Textile Laboratory' is the 3D simulator elaborated by the KAEMaRT Group, that, instead of being integrated with the designing system, is used alone, separately (because the Group is interested in producing only simulations). It also permits the user to modify the parameters associated to the various areas in which the garment has been divided.
For the simulator, instead, the Group met some problems connected with the starting physical model of reference and the solutions introduced later on permitted the construction of a more refined new model. The obtained results are still imperfect and not yet usable at the industrial level, but they demon-

strate how correct is the way that the Group is now following for the simulation, though, to be complete, it needs to undergo two further steps:

- the acquisition of experimental data deriving from the various areas in which the jacket has been subdivided;
- the optimisation of certain calculation times (a process that the Group is now working at, a part of the code has been parallelized, and they are getting good results).

During the running of the research they had to note that, with reference to the jacket, there were a series of aspects, that in the initial phase of the research had not been kept in consideration (such as the large number of layers of the material, the roll, the length, etc.). They coped with these aspects during the analysis of the modifications[17].

Bibliography

AA.VV., 'Computer Graphics', in *Proceeding Siggraph*, vol. 26, no. 2 July 1992.

AA.VV., Eurographics Association, 99.

H. Anderson (ed.), *La creatività e le sue prospettive*, La Scuola, Brescia 1975.

F. Bianchin, 'Cultural consideration in Inner City Regeneration in Council for Cultural Cooperation, Culture and Neighbour Bonds', vol. I, *Concepts and References*, Council of Europe Publishing, Strasbourg 1995.

W. Bijker and others, (eds.), *The Social Construction of Technological System. New Directions in the Sociology and History of Technology*, Mit Press, Cambridge 1987.

D. Bloor, *Knowledge and Social Imagery*, Routledge&Kegan Paul, London 1976.

D. Brandt, K. Henning, G. Strina, *Self-Similarity in Change Processes*, University Press, Aachen 1998.

A. Bruschi, *Metodologia delle scienze sociali*, B. Mondadori, Milano 1999.

17 The passage is taken from the interview released by Caterina Rizzi, Physics and KAEMaRT Senior Researcher on 01/08/2001.

F. Carignan, J. Yang, N. Magnenat Thalmann, D. Thalmann, 'Dressing animated synthetic actors with complex deformable clothes', in *Computer Graphics (Proceeding Siggraph)*, July 1992.

C. Cecchini, F. Indovina, *Simulazione*, Angeli, Milano 1989.

R. Conte, in L. Ricolfi, *La ricerca quantitativa*, Angeli, Milano 1997.

U. Cugini, M. Bordegoni, C. Rizzi, F. De Angelis, P. Prati, *Modelling and Haptic Interaction with non rigid materials*, Constraint Management, Eurographics, 1999.

F. D'Egidio, *Intuito e conoscenza*, Franco Angeli, Milano 1997

F. D'Egidio, *La fine della grande illusione*, S&K, Milano 1998.

F. D'Egidio, *La vitalità d'impresa*, Sperling & Kupfer, Milano 1999.

A. Frugoli, R. Rizzi, 'Physically-based simulation with SoftWorld', in *proceeding IASTED International Conference on Applied Modelling and Simulation*, Honolulu, August, 1998.

P. Gagliardi, *Symbols and artefacts: views of the corporate landscape*, De Guyter, New York 1990.

D. Goleman, *Lavorare con l'intelligenza emotiva*, RCS Libri, Milano 1998.

G. Holton, *L'immaginazione scientifica*, Einaudi, Milano 1973.

R. Kanter, *The change masters: Innovation and Entreprenereurship in the American Corporation*, Simon&Schuster, New York 1983.

L. Lessig, *The Future of Ideas. The Life and Death of Creativity in the Age of the Internet*, Lawrence Cloth, 2001.

T. Maldonado, *Critica della ragione informatica*, Feltrinelli, Milano 1997.

P. Medawar, *The Art of Soluble*, Methuen, London 1967.

A. Melucci, *Creatività: miti, discorsi, processi*, Feltrinelli, Milano 1994.

M. Mulkay, *La scienza e la sociologia della conoscenza*, Edizioni di Comunità, Milano 1981.

M. Negrotti, 'I paradigmi di Kuhn, la realtà e l'artificiale', in E. Campelli (ed.), *T.S. Kuhn: come mutano le idee sulla scienza*, Angeli, Milano 1999, pp. 241-268.

M. Negrotti, *Artificiale. La riproduzione della natura e le sue leggi*, Laterza, Bari 2000.

G. Natalichi, *Modelli simulativi nelle scienze sociali*, Angeli, Milano 1994.

F. Neresini, 'Creatività e contesti: relazioni e istituzioni', in A. Melucci, *Creatività: miti, discorsi, processi*, Feltrinelli, Milano 1994, pp. 191-205.

D. Parisi, *Simulazione. La realtà rifatta al computer*, Il Mulino, Bologna 2001.

A. Porrello (ed.), *Virtualità e modellistica*, IUAV, Venezia 2000.

A. Porrello, 'Information Society and the Theory of the Artificial', in M. Negrotti (ed.), *Yarbook of the Artificial – Issue 1*, Peter Lang, Bern 2002, pp. 163-181.

A. Porrello, *L'industria culturale e il prodotto moda*, Quattroventi, Urbino 2002.

J. Rex, 'The structure of Egalitarian Multi-Cultural Societies and Creative Cultural Development in Cities and Urban Neighbourhoods', presented at the meeting *Culture and Neighbour Bonds*, Council of Europe Publishing, Strasbourg 1995.

L. Ricolfi, *La ricerca quantitativa*, Angeli, Milano 1997.

C. Sander, D. Brandt, K.H. and H. Saß, *Designing and Simulating Socio-technical Systems – Concepts and Strategies*, University Press, Aachen 1998.

J. Seely, B.P. Duguid, *The Social Life of Information*, Harvard Business School Press, Boston 2000.

T.A. Stewart, *Il capitale intellettuale*, Ponte alle Grazie, Milano 1999.

T.A. Stewart, *La ricchezza del sapere. L'organizzazione del capitale intellettuale nel XXI secolo*, Ponte alle Grazie, Milano 2002.

B. Stoffels, D. Brandt, 'Software Engineering for large human-machine systems as a closed loop control process', presented at the *8th IFAC Symposium on Human-Machine Systems HMS*, Kassel, Germany 2001.

E. Taylor, J. Wheeler, *Fisica dello spazio-tempo*, Zanichelli, Bologna 1996.

G. Volino, N. Magnenat Thalmann, J. Jianhua, D. Thalmann, 'An evolving system for simulating clothes on virtual actors', in *Computer Graphics in Textiles and Apparel, IEEE Computer Graphics an Applications*, vol. 16, n. 5, September, 1999.

Antonino Porrello, IUAV, Università di Venezia

Fumihiko Satofuka

Paradigm shift of Kuhnian model to post-kuhnian one in the history of science, technology and art in Modern Japan

Abstract:

During the last decade there has been a debate on the nature of Japanese science, technology, art and their potentials. this paper addresses itself to aspects of this discussion on Japanese tradition, locating it in a larger civilizational context and posing questions concerning the social conditions of Japanese tradition. My paper covers the basic outline of the paradigm shift of Kunian Models (Fission Model) to post-kuhnian ones (Fusion Model) in science, technology and art.

Keywords:
science, technology, art, culture, tradition, fusion, fission

1 Science and Culture in Japan in Comparison with the Tradition of Art and Technology

Japanese scientific tradition has taken a multifaceted approach.

It constitutes a carefully directed effort. The suggested changes include a move away from rote learning, the removal of the hierarchical order in school, changes in organizational structures of laboratories in favour of those promoting innovation, increases in funding, expanding links between industry and research and the careful cultivation of foreign scientist and foreign journals.

A formidable attempt seems to be under way. The question we should pose is, will this succeed?

This question was solved by using the kuhnian Model (Fission Model).

One of the elements behind the lack of productivity and tradition of science, technology and art in culturally dependent countries that has been identified is the incomplete and fragmentary nature of the mapping of 'international' science.

It has been documented that most of the time the different disciplines and sub-disciplines are incompletely mapped through teaching programs, journals, reading material and laboratories in culturally dependent countries.

In the case of Japan, however, this is not the case. The mapping is as complete as in any Western country and its science, technology and art are fully integrated with those of the West. These sources of the lack of productivity do not exist in the case of Japan.

2 Artificial Fusion

Critics and commentators in the theatre, dance and music fields have recently observed significant elements of traditions.

In Japan today, a dominant position is held by Western music. This is not only so in the case of pop music but is perhaps more so in serious music.

Serious, classical music in current Japan. It appears, is not the music of its tradition but Western music. The first landmark event in the history of contemporary music in Japan was in 1952 when Matsudaira Yoritsune won the first prize for a composition 'Theme and Variation for Piano and Orchestra'.

This piece combined twelve tone techniques with a traditional genre of Japanese music, namely *'gagaku'* in the late 1950s, many Japanese felt that they had 'caught up' with the standards of European music, as exemplified by Mayuzumi's Nirvana Symphony. This composition, it should be noted, employed Buddhist temple bells and Buddhist chanting. In the 1960s and 1970s traditional Japanese musicians began to perform Western-style composition using traditional instruments.

Thus these trends were characterized by the post-Kuhnian Model (Fusion Model).

3 Technological Fusion

There have been several important examples of such fusion from such fission.

They are in those Japanese technologies which are identified with Japanese innovation, namely, robotics, biotechnology and the new ceramics.

This interdisciplinary technological strategy combines two existing areas of knowledge to give new products and new industries. The borders between such two disciplines are now blurred and many innovations may emerge by combining the two.

Here no 'fundamental' creativity is required, but an imaginative application of existing knowledge at a higher level of integration. Such kind of trend was characterized by the paradigm shift of the Kuhnian Model to post-Kuhnian Model (Fission to Fusion). In this technological fusion, two basic technologies are combined into one to yield a new field of technology.

The classic example is the Japanese experience in *mechatronics* – a field vital for robots – which combined mechanics and electronics. *Mechatronics* is a Japanese innovation where mechanical technology was fused with electronic technology.

Examples of *mechatronics* are in Numerically Controlled (NC) machine tools and robots. This category also includes technologies where the mechanical working part is either wholly or partly superseded by electronics. A *mechatronized* process it should be noted, was influenced by a MITI initiative resulting in a law in 1971 showing that direct legislation can have an influence on technological innovation. The innovation sequence in product fields other than its principal product line. In *mechatronics*, technological fusion commenced after 1971 with a crossflow of ordinary machinery industry and the electrical machinery industry. Later, connections were created between precision instruments and communication and electronic equipment. With these four elements under fusion the new field was born.

4 Scientific Fusion

Whilst having noted the possibilities for tradition that have been described in the page s above, one should remember that cultural tradition, at least in science, had existed in Japan from the *Meiji* period onward. In spite of a blanket of conformity, there have been several examples of Japanese tradition in science going back to the early period of its introduction.

For example, Kitasato Shibasaburo discovered natural immunity in medicine, although the 1901 Nobel Prize for this was given to Von Behring of Germany alone. Similarly, during World War I, Yamagiwa Katsusaburo of Tokyo University discovered a technique for inducing tumors into laboratory animals. It is true that some of this work was done abroad – Shibasaburo did his work in collaboration with Von Bering.

A more recent examples of Japanese creativity would be Kimura's neutral theory of evolution which posits that evolution did not take place primarily through competition. Kimura has suggested that evolution proceeds not only by competition but also by adaptively neural mutation. This is a major departure from the 'struggle for survival' model of evolution going back to Darwin.

Although research has to be done before any firm statement can be made, it is perhaps suggestive that Kimura mapped-consciously or unconsciously-some elements of Japanese-type group relations and 'wa' in his model, as Darwin did the opposite in his model, illustrating the creative potential of cultural crossflows.

This trend of the scientific discoveries was characterized by the paradigm shift of the Kuhnian Models to the post-Kuhnian ones (Fission to Fusion).

5 Conclusion

The Japanese attempts at creating a new scientific tradition take place within a particular national setting. This social structure had developed out of *pre-Meiji* origins under criteria, carefully laid down from to time, by members of the Japanese ruling elite. These efforts had given rise to a highly conformist, though very productive society. It had also attendant negative social facets.

It is very unlikely therefore that these holders of power would allow the traditional slogan, updated *'wakon Yosai'* (Japanese Spirit, Western Civilization) to go beyond defined bounds so as to threaten the status quo.

Thus, the above trends in Japanese Science, Technology and Art were characterized by the paradigm shift from the Kuhnian Models (Fission Models) to the post-Kuhnian ones (Fusion Models).

Bibliography

T. Morris-Suzuki, *The Technological Transformation of Japan: From Seventeenth to the Twenty-first Century*, Cambridge University Press, Cambridge 1994.

S. Nakayama, *Science, Technology and Society in Post-War Japan*, Kegan Paul International, Tokyo 1991.

T. Hayashi, *The Japanese Experience in Technology – from Transfer to Self-Reliance*, United Nations University Press, Tokyo 1990.

I. Inkster, F. Satofuka, *Culture and Technology in Modern Japan*, I.B. Tauris, London 2000.

Fumihiko Satofuka, Department of the Environment and Creation, Daitobunka University

Andrea Sormano

Max Weber's Methodological Individualism as 'Sociological Grammar'

Abstract:

Starting from C.W. Mills original reading of Weber's actor's 'motives' in a grammatical key, I am here proposing a wider comparison between the individualistic model Weber formulates for sociology and the grammatical model Wittgenstein formulates for philosophy and reaching, in the end, the conclusion that Weber may be seen as the founder of a sociological grammar wherein motives become the meeting point for meaning, identity and power. As a model of analysis for reality, weberian sociological grammar works at the same time as a model of analysis for merely imaginary representations, such as idols and monsters. The connection between meaning, identity and power assumed as an inseparable unit, working towards establishing the sense of social actions, whether it be a real or imaginary one, as rooted deeply in language, is a sociological fall-out which the sociologist looking for new sources and foundations may gain today from reading Wittgenstein's philosopy as a chance for re-reading Weber's sociology.

Keywords:
methodological individualism, philosophical grammar, vocabularies of motive, language and culture, texts and contexts, ideal types, meaning, identity, power, real and imaginary, mytology, idols, monsters

1 Introduction

A number of sociologists, no matter what their sociological field may be, have for quite a while now turned to Wittgensteins' 'philosophical grammar', i.e. the 'grammatical' model harbouring Wittgenstein's so-called second phase, as to a main resource for renewing or refounding their own specialty. This holds true as far as the 'soft' sociological specialties, such as the sociology of knowledge, science and culture, as well as the sociology of language, are concerned. However, most remarkable among recent trends, is the fact that this holds true in the case of 'hard' sociological fields as well. In other words, Wittgenstein's notions of 'grammar' and 'language-game' have brought about new and surprising approaches in such specialties as the sociology of work, industry and organization[1]. Assuming so novel a tendency as a background I will try to probe into one of its facets cerning up occasionally or in an indirect way, namely, the kind of relation one can establish between the old and the new, between Weber and Wittgenstein. My essay will try to find out what kind of relation one can ascertain between the former's 'individualistic' method of analysing social actors' *motives* on the one side, and the latter's 'grammatical' method of observing *language-games* on the other. I will take the lead from C.W. Mills's *Situated Actions and Vocabularies of Motive*, the essay in which he lays the foundations for such comparative perspective.

2 An analytic model for the explanation of motives

C.W. Mills was the first to suggest 'an analytic model for the explanation of motives' in sociology, in his early essay, *Situated Actions and Vocabularies of Motive* (Mills 1940). What I will now focus on within this 'model'[2] is not so much its forward-looking standpoint, namely its anticipating Austin's *speech act theory*[3] or Wittgenstein's *philosophi-*

1 See, among others, Coulter's contribution to founding a new sociology of knowledge, seen as a 'socio-logic' for the first group. For the second group, see Pentland's contribution to renewing the sociology of organization by formulating a 'theory of organizational knowledge' (Pentland 1992, Pentland and Rueter 1994). See Maynard (1991) and Brown (1990) for a critical assessment of the linguistic and rhetorical turn in sociology.

2 In this essay, Mills uses the word 'model' without further qualifying it; he supplies useful directions about it in his previous essay *Language, Logic, and Culture* (1939), I am referring to here.

3 Cf. Scott and Lyman (1968).

cal grammar[4], but rather its retrospective far-sightedness in his crediting Weber with being the first to construct a linguistic and cultural (con-textual) model to analyse an actor's *motives*. In the opening sentences of the essay, Mills writes:

> The postulate underlying the modern study of language is the simple one that we must approach linguistic behavior, not by referring to private states in individuals, but by observing its social function of coordinating diverse actions. Rather than expressing something which is prior and in person, language is taken by other persons as an indicator of future actions (Ibid. p. 904).

Starting from this perspective, Mills solves the 'problems of motivation' abandoning the inferential conception that assumes motives as subjective 'springs' of action and drawing up an 'analytic model for the explanation of motives' within which motives acquire the public identity of 'typical vocabularies having ascertainable functions in delimited societal situations'. To explain behavior by referring it to an inferred and abstract 'motive' is one thing, according to Mills; to analyze the observable 'lingual mechanism' of motive imputation and avowal as they function in conduct is quite another. Rather than fixed elements 'in' an individual, this second perspective takes motives to be 'the terms in which interpretation of conduct by *social actors* proceeds'. This imputation and avowal are phenomena to be explained, which Mills goes about doing by establishing what are the general conditions under which such motive imputation and avowal seem to occur.

Mills argues that one rarely questions one's motives for action; one does it however, typically, when '*alternative* or *unexpected* programs of actions' are present, that is to say, when critical situations alter a course of action which otherwise would not raise any 'question'. Men live in immediate acts of experience and their attention is directed outward until acts are in some way frustrated: it is then that awareness of self and motive occur. Motives are 'words' imputed or avowed as answers to questions interrupting acts or programs. They do not denote any elements 'in' individuals, but rather stand for 'anticipated situational consequences of questioned conduct': intention or purpose

4 Cf. Blum and McHugh (1971), Coulter (1989).

is awareness of anticipated consequences, motives are *names* for consequential situations, and surrogates for actions leading to them. On the basis of such considerations, Mills criticizes Dewey's 'model of purposive conduct', according to which individuals confronted with 'alternative acts' perform either one on the basis of differential consequences which they anticipate. This 'nakedly utilitarian schema' is inadequate, according to Mills, because it does not reckon, first of all, with the fact that 'alternative acts' of *social* conduct appear most often in *lingual form*, namely as a question posed to oneself or to somebody else, and secondly that individuals act in terms of anticipation of *named* consequences. A sociological conception of motives as 'relatively stable lingual phases of delimitated situations' is on the contrary quite consistent, Mills goes on to maintain, with Mead's program designed to approach conduct socially and from the outside: it translates the question of 'why' into a 'how', that is answerable 'in terms of a situation and its typical vocabulary of motives, i.e., those which conventionally accompany that type situation and function as cues and justifications for normative actions in it' (Ibid. p. 906).

The point I want to stress is that, according to Mills, this very 'sociological conception of motives' is to be found in Max Weber's thought and proves central to *Economy and Society* (Cap. I, § 7), where Weber defines motive as 'a complex of meaning, which appears to the actor himself or to the observer to be an adequate ground for his conduct'. The aspect of motive which this conception grasps, Mills argues, is its 'intrinsically social character':

> A satisfactory or adequate motive is one that satisfies the questioners of an act or program, whether it be the other's or the actor's. As a word, a motive tends to be one which is to the actor and to the other members of a situation an unquestioned answer to questions concerning social and lingual conduct. A stable motive is an ultimate in justificatory conversation. The words which in a type situation will fulfil this function are circumscribed by the vocabulary of motives acceptable for such situations. Motives are accepted justifications for present, future, or past programs or acts (Mills 1940, p. 907).

In the remaining part of his essay Mills further develops his 'analytic model', without mentioning either Weber or, even less, Wittgenstein, whose philosophical grammar he could not have known since its author was still working on it. I will try to do it in what follows, confining myself to pointing up those elements in such model which can more strictly be connected with Weber's methodological individualism and with Wittgenstein's philosophical grammar[5].

When an agent vocalizes or imputes motives, Mills argues, anticipating a distinction Austin could have made, he is not trying to 'describe' his experienced social action, nor is he merely stating 'reasons', but he is 'influencing others – and himself'. Thus, firstly, we need not treat an action, as discrepant from 'its' verbalization, for in many cases 'the verbalization is a new act'. In such cases, there is no discrepancy between an act and 'its' verbalization, but a difference 'between two disparate actions, motor-social and verbal' (Ibid., p. 907). Secondly, the sociological truth of the motive, cannot be looked for in the 'sincerity' of those pronouncing them, but in its power to solve the crisis or the conflict which lies at the basis of its occurrence. Since motives are communicated, they may be lies, but such lack of sincerity may be quite irrelevant if it is not relevant to the addressee. If we translate this in Wittgenstein's terms, the meaning of a motive cannot be looked for in the truth of sensation or feeling as experienced by the one who expresses them, but are to be looked for in the 'game' in which the motive occurs[6]. The sense of Mills's motives is like the sense of a proposition: it does not rely on the 'soul' or 'atmosphere' that may go along with it, rather it depends on the circumstances of its usage[7]. If we go back to Weber's thought we will

5 On the connection between Weber and Wittgenstein (and Benveniste) I extensively wrote in my previous work (Sormano 2000 and 2003).

6 'We may say that the words "fine", "oh", and also "perhaps" are *expressions* of sensation, of feeling. But I don't call the feeling the meaning of the word. We are not interested in the relation of the words to the sensation, whatever it may be, whether they are evoked by it, or are regularly accompanied by it, or give it an outlet. We are not interested in any empirical facts about language, considered as empirical facts. We are only concerned with the description of what happens and it is not the truth but the form of the description that interests us. What happens considered as a game' (PG, 30).

7 'The sense of a proposition is not a soul. It is only in a language that something is a proposition. To understand a proposition is to understand a language. A proposition is a sign in a system of signs. It is *one* combination of signs among a number of possible ones, and as opposed to other possible ones.' (PG 84d) 'You say to

find that motive, as well as power, is intrinsically relational, it becomes real (satisfactory, adequate, unquestioned answer) only when faced with somebody's legitimation, no matter what his private 'truth' may be.

Farther on in Mill's essay, further analogies with both Wittgenstein's philosophy and Weber's sociology come to the fore through a definite counter-psychological strain:

> Generally, motives are imputed by others before they are avowed by self. The mother controls the child: 'Do not do that, it is greedy.' Not only does the child learn what to do, but he is given standardized motives which promote prescribed actions and dissuade those proscribed. Along with rules and norms of action for various situations, we learn vocabularies of motives appropriated to them. These are the motives we shall use, since they are a part of our language and components of our behavior (Ibid., p. 909).

What Mills shows us in these passages corresponds to an example of Wittgenstein's language-games, i.e. one of the 'countless' examples of the fact that 'the *speaking* of language is part of an activity, or of a form of life'[8]. If it does not go along with an act of signification, an activity is no 'rule-guided' *action*, as Blum and McHugh maintain, but it is a 'rule governed' *behaviour*, i.e., it is not a 'meaningful action' as Weber would have defined it, but a 'merely reactive behaviour' (E§S, 4)[9]. In Mills's view, motives are not primarily real when they

---

me: "You understand this expression, don't you? Well then – I am using it in the sense you are familiar with." – As if the sense were an atmosphere accompanying the word, which it carried with it into every kind of application. If, for exemple, someone says that the sentence "This is here" (saying which he points to an object in front of him) makes sense to him, then he should ask himself in what special circumstances this sentence is actually used. There it does make sense' (PI, 117).

8 'But how many kinds of sentence are there? Say assertion, question, and command? – There are *countless* kinds: countless different kinds of use of what we call "symbols", "words", "sentences". And this multiplicity is not something fixed, given once for all; but new types of language, new language-games, as we may say, come into existence, and others become obsolete and get forgotten. [...] Here the term "language-*game*" is meant to bring into prominence the fact that the *speaking* of language is part of an activity, or of a form of life.' [...] (PI, 23).

9 In 'The Social Ascription of Motives', Blum and McHugh (1971) use Weber's distinction between 'meaningful action' and 'merely reactive behavior' and un-

are sincere, but when they are apt to solve the crisis which determines their occurrence. As such, real motives cannot be looked for in a metaphysical way, beyond the surface of their verbalization in the realm of biology or psychology. Mills does not oppose 'real motives' to 'mere rationalization', nor does he agree with the view held by many sociologists that language 'is an external manifestation or concomitant with something prior, more genuine, and "deep" in the individual'.

---

derline the role rules and distinctions among rules play in founding actions as *meaningful* activity. 'Rules make actor's methodicity and concern possible by transforming what would otherwise be nonsense into intelligible social behavior. That alter can be conceived (by observing ego and sociologist) to be rule-guided encompasses the status of motive within general sociology: motive, as with any other sociological classification refers to certain actions by rule-guided ascribers and objects. [...] For motive treatment to occur, as for any kind of membership treatment to occur, ego and alter must necessarily generate for one another their status as members oriented to rules. Here a set of concrete activities is so formulated that those activities become members through the application of a corpus of rules. This is a detailed way of saying they are members of a social relation [...] Take, for example, the brute. There are things which can be said of such an organism, *e.g.*, it is enraged, it is contented, it is sleeping, and so forth. There are characterizations, in that they depict some state of affairs, and rule-guided ascriptions at that. But they are not, and could not be, common-sense ascriptions of motive because the brute's behaviors are not thought to be displays by a rule-guided actor – the brute is thought not to be socially "responsible". He is thought not to be responsible insofar as his activity is, in Weber's terms, behavior rather than action. Nor is the brute thought to be a bona fide member, and for the same reason: he cannot be a member because he cannot be said to know what he's doing through an orientation to member rules. [...] The distinction between rule-governed and rule-guided is comparable to the one between behavior and action, and it is surely correct that the use of rules by members is an accomplishment in the hardiest sense of the term' (Ibid., p. 104-105). Blum and McHugh's analysis of motives, which I am restricting to a single passage, does not seem to provide an alternative to Mill's own as its authors claim; rather they seem to point to a different and complementary level of analysis. Such analysis concerns the criteria by which an actor or an observer confer motive identity to an activity, part of which covers the above mentioned difference between behaviour and action. Mills does not shut off this level of analysis, on the contrary he insists on it in the last passages mentioned above; he is nevertheless interested mainly in motives in action, so to speak, that is to say in the circumstaces of their occurrence, in the contextual criteria an actor uses in choosing them, in their function in orientating courses of action.

Distinction between real attitudes and mere verbalization or opinion implies, according to Mills, that at best we only infer from his language what 'really' an individual's attitude or motive is. But what could we possibly so infer? Of precisely what is verbalization symptomatic?

> We cannot infer physiological processes from lingual phenomena. All we can infer and empirically check is another verbalization of the agent's which we believe was orienting and controlling behavior at the time the act was performed. The only social items that can 'lie deeper' are other lingual forms. The 'Real Attitude or Motive' is not something different in kind from the verbalization or the 'opinion'. They turn out to be only relatively and temporally different. [...] Within the perspective under consideration, the verbalized motive is not used as an index of something in the individual but as a basis of inference for a typal vocabulary of motives of a situated action. [...] There is no way to plumb behind verbalization into an individual and directly check our motive-mongering, but there is an empirical way in which we can guide and limit, in given historical situations, investigations of motives. This is by the construction of typal vocabularies of motives that are extant in types of situations and actions. Imputation of motives may be controlled by reference to the typical constellation of motives which are observed to be societally linked with classes of situated actions. [...] The only source for a terminology of motives is the vocabularies of motives actually and usually verbalized by actors in specific situations (Mills 1940, p. 909-910).

Seen from a Weberian point of view, what Mills is saying is that we are not to look for 'the ultimate foundation of the sociological interpretation of action' among the 'laws' of psychology, but rather among the rules of culture (and of language) that publicly (idealtypically, grammatically) make the actor's motives intelligible, whether they be rational or irrational[10]. Like Wittgenstein, Mills urges us not to try to

10 It is erroneous, Weber writes in *Economy and Society*, 'to regard any kind of psycology as the ultimate foundation of the sociological interpretation of action. [...] The source of error lies in the concept of "psychic". It is held that everything which is not physical is *ipso facto* psychic. However, the *meaning* of a train of mathematical reasoning which a person carries out is not in the relevant sense "psychic" [...] On the other hand, in explaining the irrationalities of action sociologically, that form of psycology which employs the method of subjective understanding undoubtedly can make decisively important contributions. But this does not alter the fundamental methodological situation' (E&S, 19).

'penetrate' motives, in other words, not to look through motives for something that lies beyond the verbalizations motives are made of[11], but to assume the very motives as 'proto-phenomenonon' or as a '*primary* thing' to be described according to the place they hold within the grammar, and not to be interpreted according to our experiences and personal feelings[12]. How do we go about it? We are to employ those grammatical standards of measurement that Weber defines 'ideal types', Mills names 'typical vocabularies of motives', and Wittgenstein 'language games'. For the three of them the problem of understanding action is solved in a counter-psychological way by using everyday language grammar as an analytical, model of internal connections and oppositions.

3 Weber and Wittgenstein: a comparison.

According to Wittgenstein, grammar 'tells what kind of object anything is' (PI, 373). In Wittgenstein's second philosophical phase, grammar replaces ontology by solving ontological determinations turning them into grammatical determinations. Any time we seem to have run into a necessary aspect of the 'world', we are actually confronted with a grammatical connection, i.e. with connections between *concepts*, and we are dealing with aspects of language within which we *speak* about the world. A conceptual investigation is therefore a

11 'We feel as if we had to *penetrate* phenomena: our investigation, however, is directed not towards phenomena, but, as one might say, towards the *"possibilities"* of phenomena. We remind ourselves, that is to say, of the *kind of statement* that we make about phenomena […] Our investigation is therefore a grammatical one […]' (PI, 90) 'One ought to ask, not what images are or what happens when one imagines anything, but how the word "imagination" is used.' (PI, 370).

12 'Our mistake is to look for an explanation where we ought to look at what happens as a "proto-phenomenon". That is, where we ought to have said: *this language-game is played*' (PI, 654). 'The question is not one of explaining a language-game by means of our experiences, but of noting a language-game' (PI, 655). 'What is the purpose of telling someone that a time ago I had such-and-such a wish? – Look on the language-game as the *primary* thing. And look on the feeling, etc., as you look on a way of regarding the language-game, as interpretation […]' (PI, 656).

necessary prerequisite in order to avoid any metaphysical lack of distinction between factual and conceptual investigations[13]. But once we enter into conceptual investigations, the philosopher's grammatical (anti-metaphysical) task consists in ascertaining and separating the respective domains of possibility and necessity:

> When one wants to show the senselessness of metaphysical turns of phrase, one often says 'I couldn't imagine the opposite of that', or 'What would it be like if it were otherwise?' [...] Well, if I can't imagine how it might be otherwise, I equally can't imagine that it is *so* (PG, 83b).

If negating a statement is meaningless, then the very statement is meaningless. On such semantical premise (the 'cornerstone of Wittgenstein's thought'[14]), he bases his conviction that the facts of the world are contingent[15] and that the only kind of necessity concerns language rules. Moreover, we are not dealing here with a kind of necessity enforced by any other reality than the 'arbitrary' one belonging to language itself, the one which enables us to think, to express ourselves, to describe reality[16]. In as much as it is a descriptive principle, by which we use words within language[17], grammar allows us to spot the very possibilities of combination and opposition of which language consists. Seen from this point of view, the goal of philosophy lies no longer in seeking the 'essence' of language, as it was the case

13 'Philosophical investigations: conceptual investigations. The essential thing about metaphysics: it obliterates the distinction between factual and conceptual investigations' (Z, 458). See Winch (1958) on the relevance of this distinction to sociological investigation.
14 Cf. Marconi (1997, p. 78).
15 If there were necessary facts their descriptions would consist of statements the negation of which would be meaningless ('I cannot imagine the opposite': negating a necessary fact is unthinkable); then the very descriptions would be meaningless, which equals saying that necessary facts are unthinkable. Cf. Marconi (1997, p. 78, n. 31).
16 'The rules of grammar may be called "arbitrary", if that is to mean that the *aim* of the grammar is nothing but that of the language' (PI, 497). See also PG, 140b.
17 'Grammar describes the use of words in the language. So it has somewhat the same relation to the language as the description of a game, the rules of a game, have to the game' (PG, 23). 'The question "What is a word really?" is analogous to "What is a piece in chess?"' (PI, 108).

for the author of *Tractatus Logico-Philosophicus*[18], but in throwing light 'on the facts of our language', and producing just that understanding which consists in 'seeing connections'[19] by turning to language-games as 'models', i.e. as 'objects of comparison'[20]. Having been thus transformed, philosophy replaces 'wild conjectures and explanations by quiet weighing of linguistic facts' (Z, 447): rather than a language theory, philosophy becomes a 'therapy'[21]. Having left behind his essentialist prejudice, the philosopher-therapist 'treats' language problems the way one treats illness[22], in other words he describes the reality of language usages and along with it he recognizes and describes the 'entire mythology' laid down in our language[23].

18 'It is interesting to compare the multiplicity of the tools in language and of the way they are used, the multiplicity of kinds of words and sentences, with what logicians have said about the structure of language. (Including the author of the *Tractatus logico-Philosophicus*)' (PI, 23). 'When philosophers use a word – "knowledge", "being", "object", "I", "proposition", "name" – and try to grasp the *essence* of the thing, one must always ask oneself: is the word ever actually used in this way in the language-game which is its original home? – What *we* do is to bring words back from their metaphysical to their everyday use' (PI, 116).

19 'A main source of our failure to understand is that we do not *command a clear view* of the use of our words. – Our grammar is lacking in this sort of perspicuity. A perspicuous presentation produces just that understanding which consists in "seeing connections". Hence the importance of finding and inventing *intermediate cases*' (PI, 122). See also RFGB, pp. 69-70.

20 'Our clear and simple language-games are not preparatory studies for a future regularization of language – as it were first approximations, ignoring friction and air-resistance. The language-games are rather set up as *objects of comparison* which are meant to throw light on the facts of our language by way not only of similarities, but also of dissimiliarities' (PI, 130). 'For we can avoid ineptness or emptiness in our assertions only by presenting the model as what it is, as an object of comparison – as, so to speak, a measuring-rod; not as a preconceived idea to wich reality *must* correspond. (The dogmatism into which we fall so easily in doing philosopy)' (PI, 131).

21 'There is not *a* philosophical method, though there are indeed methods, like different therapies' (PI, 133).

22 'The philosopher's treatment of a question is like the treatment of an illness' (PI, 255). 'What is your aim in philosophy? – To shew the fly the way out of the fly-bottle' (PI, 309).

23 'An entire mythology is laid down in our language' (P, 93).

Such statement as 'grammar tells what kind of object anything is' can be easily attributed to Weber if we follow Mills's precise hints in that direction. I will therefore assume Weber's 'science of reality' as a grammar of motives belonging to the social actor as a 'cultural being'. Weber decrees the end of historicism as a consequence of the end of ontology[24]. Of course Weber would not maintain that history is at an end: he would rather criticize any idea of history taken as a reality ontologically separate and independent from social actors' stances, who are in the end history's only authors. This position does away with a linial view of history, whether it be a positivistic ideology of progress or a spiritualistic metaphysics. It does away with mingling ethical evolutionism and historical relativism, a mixture which is inconsistent with any perspective of social science whatsoever. From Weber's standpoint, history becomes a 'possibility' and, as such, the special subject of a science of reality (as opposed to a 'science of laws'), the reality of which is a 'culture' which Weber assumes and assesses in a way resembling the one Wittgenstein uses to assume and assess language, that is, as a system of combinations and oppositions, of 'elective affinities' and 'death-struggles', of value free enunciations as opposed to value judgements. In Weber's view too, a historical fact or an actor's motive, a personal identity or a kind of power, are each and all liable to be described as such – are *so* – only provided that their cultural meaning be assumed as inevitably connected whith their *otherwise* (Meth., p. 72), that is, located within a system of conventional rules as opposed to a set of natural laws. His ideal types taken as 'conceptual instruments for *comparison* with and the *measurement* of reality' (Ibid., p. 97) are as fundamental to such ways of describing as Wittgenstein's language-games, to which they are strictly connected. In both thinkers such a model is liable to describe both the reality of culture in action and the entire mythology so widely laid down within it.

The 'historical human being', Weber writes in his *Objective possibility and adequate causation in historical explanation*, 'has an attitude and will of his own', and 'would never "act" if his own action appeared to him as "necessary" and not only as "possible"' (Meth.,

24 Cf. Tessitore (1996).

p. 165). Precisely on this possibility can an empirical science work: 'An empirical science', Weber writes in *'Objectivity' in Social Science and Social Policy*, 'cannot tell anyone what he *should* do – but rather what he *can* do – and under certain circumstances – what he wishes to do' (Ibid., p. 54). An empirical science having taken on the function of 'cultural science' fulfils such task, a therapeutic task according to Wittgenstein' sense, if it solves what look like imaginary ontological determinations or objective realitites turning them into real individual and cultural determinations[25]. 'The trascendental presupposition' of a cultural science seen as *'empirical science* of concrete *reality'* (Ibid., p. 72), when we distinguish it from 'sciences constituted by laws' (Ibid., p. 133),

> lies not so much in our finding a certain culture or any 'culture' in general to be *valuable* but rather in the fact we are *cultural beings*, endowed with the capacity and the will to take a deliberate attitude towards the world and to lend it *significance* (Ibid., p. 81).

A cultural scientist can help in specifying what one 'can' do and 'wishes' to do, in so far as he offers criteria to ascertain 'the significance' of what one wants; in other words, in so far as he finds out the conceptual status of the ideas that lie at the basis of such and such a goal by pinpointing each idea within its cultural structure of connection or opposition between ideas, and within the context in which it occurs[26].

25 'It is true that we regard as *objectively* valuable those innermost elements of the "personality", those highest and most ultimate value-judgments which determine our conduct and give meaning and significance to our life' (Meth., p. 55).

26 'Science can make him (the acting, willing person) realize that all action and naturally, according to the circumstances, inaction imply in their consequences the espousal of certain values – and herewith – what is today so willingly overlooked – the rejection of certain others. The act of choice itself is his own responsibility. We can also offer the person, who makes a choice, insight into the significance of the desired object. We can teach him to think in terms of the context and the meaning of the ends he desires, and among which he chooses. We do this through making explicit and developing in a logically consistent manner the "ideas" which actually do or which can underlie the concrete end. It is self-evident that one of the most important tasks of every science of cultural life is to arrive at a rational understanding of these "ideas" for which men either really or allegedly struggle.

In Weber's view, empirical reality is something only in so fas as it is '*significant* to *us*', i.e. 'is related to the *cultural values* with which we approach reality' (Ibid., p. 78). Significant to us is always 'a finite part alone of the infinite variety of phenomena', and on the basis of such presupposition only 'the knowledge of an individual phenomenon becomes logically meaningful'. An 'objective' analysys of cultural events, which proceeds according to the thesis that the ideal of science is the reduction of empirical reality of 'laws' is, according to him, 'meaningless':

> It is not meaningless, as is often maintained, because cultural or psychic events for instance are 'objectively' less governed by laws. It is meaningless for a number of other reasons. Firstly, because the knowledge of social laws is not knowledge of social reality but is rather one of the various aids used by our minds for attaining this end; secondly, because knowledge of cultural events is inconceivable except on a basis of the significance which the concrete constellations of reality have for us in certain individual concrete situations. In which sense and in which situations this is the case is not revealed to us by any law; it is decided according to the value-ideas in the light of which we view 'culture' in each individual case. 'Culture' is a finite segment of the meaningless infinity of the world process, a segment on wich human beings

---

This does not overstep the boundaries of a science which strives for an "analytical ordering of empirical reality", although the methods which are used in this interpretation of cultural values are not "inductions" in the usual term. [...] But the scientific treatment of value-judgments may not only understand and empathically analyze the desired ends and the ideals which underlie them; it can also "judge" them critically. This criticism can of course have only a dialetical character, i.e., it can be no more than a formal logical judgment of historically given value-judgments and ideas, a testing of the ideals according to the postulate of the internal *consistency* of the desired end. It can, insofar is it sets itself this goal, aid the acting willing person in attaining self-clarification concerning the final axioms from which his desired ends are derived. It can assist him in becoming aware of the ultimate standards of value which he does not make explicit to himself or, which he must presuppose in order to be logical. The elevation of these ultimate standards, which are manifested in concrete value-judgments, to the level of explicitness is the utmost that the scientific treatment of value-judgments can do without entering into the realm of speculation. As to whether the person expressing these value-judgments *should* adhere to these ultimate standards is his personal affair; it involves will and conscience, not empirical knowledge' (Meth., pp. 53-54).

> confer meaning and significance. This is true even for the human being who views a particular culture as a mortal enemy and who seeks to 'return to nature'. He can attain this point of view only after viewing the culture in which he lives from the standpoint of his values, and finding it 'too soft'. This is the purely logical-formal fact which is involved when we speak of the logically necessary rootedness of all historical entities (historische Individuen) in 'evaluative ideas' (Ibid., pp. 80-81).

Having assumed that the reality of a science of reality is culture as we defined it above, Weber goes on to locate and measure its components by turning to those 'analytical constructs' *(Gedankenbilder)* which he calls 'ideal types'. An ideal type is an 'ideal picture', a 'utopia', a 'conceptual pattern' conceived as 'an internally consistent system'[27]; it is neither (historical nor sociological)[28] reality, nor even the 'true' reality, nor a 'description' of reality, but rather 'it aims to give unambiguous means of expression to such a description' (Meth., p. 90). Ideal type is not a 'model' of what 'ought' to exist, but 'an analytical construct, which is "ideal" in the strictly logical sense of the term' (Ibid., p. 92); it is fitted to serve 'as a schema under which a real situation or action is to be subsumed as one *instance*'; it has the significance of 'a purely ideal *limiting* concept with which the real situation or action is *compared* and surveyed for the explanation of certain of its significant components' (Ibid., p. 93); it is a conceptual instrument 'for *comparison* with and the *measurement* of reality' (Ibid., p. 97). Weber's ideal type is a 'model' in the way Wittgenstein's language-game is a model: an 'object of comparison', not a 'preconceived idea to which reality *must* correspond'[29]. Through this model both avoid

27 'An ideal type is formed by the one-sided *accentuation* of one or more points of view and by the synthesis of a great many diffuse, discrete more or less present and occasionally absent *concrete individual* phenomena, which are arranged according to those one-sidedly emphasized viewpoints into a unified *analytical* construct. In its conceptual purity, this mental construct cannot be found empirically anywhere in reality. It is *utopia*' (Meth., p. 90).

28 On the distinction between olistic (historical) and individualistic (sociological) ideal types see Watkins (1952). For a critique of such distinction Cohen (1966). On ideal types as elements of a 'socio-historical model' see Roth (1975, 1976).

29 Cf. n. 20.

the dogmatism 'into which we fall so easily in doing philosophy'[30] or sociology.

Ideal types for Weber are 'ideal' in two respects: on the one hand, they are always based on a concept of logical and ideational perfection and they pursue this, through many considerations, to a conceivable extreme; on the other hand, they are also related to 'ideas' i.e. they are 'analytical constructs' *(Gedankenbilder)*, thus they are plans for thought *(Gedanken)*[31]. The task of the ideal type is to set a logical and ideational order in the chaotic multiplicity of individual phenomena – in the 'meaningless infinity of the world process' – and as such, ideal types too are infinite. Nothing is farther from Weber's position than planning to define ideal types as unifying concepts, or conceiving an inventory of ideal types. The fact that reality never comes into consideration when ideal types are involved, makes them an 'exclusively formal instrument for the intersubjective, discursive understanding of historical reality'[32]. The very 'discursive nature of our knowledge', according to Weber, postulates ideal types as 'a conceptual shorthand'[33], as an essential tool to our 'nomological' knowledge when applied to language and cultural rules[34]. As such, ideal types are 'simply unavoidable', and the only question is 'how and for what purpose' should any particular type be constructed[35], as well as what makes the difference between using them deliberately and ignoring their funcion. If one does not accept them for what they are, i.e. tools intended for

30 Ibid.

31 Cf. Käsler (1979 [1988], p. 182).

32 Ibid.

33 'However the concepts thereupon become ideal-*typical* in the sense that they appear in full conceptual *integrity* either not at all or only in individual instances. Here as elsewhere every concept which is not purely classificatory diverges from reality. But the discursive nature of our knowledge, i.e., the fact that we comprehend reality only through a chain of intellectual modifications postulates such a conceptual short-hand' (Meth., p. 94).

34 '[...] a leading philologist today declares that the subject-matter of philology is the "speech of *every individual*", even the formulation of such a program is possible only after there is a relatively clear ideal type of the written language, which the otherwise entirely orientationless and unbounded investigation of the infinite variety of *speech* can utilize (at least tacitly)' (Meth., p. 104).

35 Cf. Albrow (1990, p. 151).

measuring cultural reality, one does not really dispense with them but, whether one realizes it or not, one ends up using similar concepts only 'without formulating them verbally or elaborating them logically' or otherwise 'remains stuck in the realm of the vaguely "felt"' (Meth., p. 94). Following such premises, Weber's individualistic method can be seen as a way to formulate verbally and to elaborate logically the imaginary side of our knowledge: it is not against studying communities, but it objects to using 'indifferentiated collective concepts'[36], to relying on *Weltanschauungen*[37], to having recourse to 'artistic portrayal' of the 'purely direct concrete' descriptions of reality ('Each sees what is in his own heart') regardless of the fact that 'valid *judgments* always presuppose the *logical* analysis of what is concretely and immediately perceived, i.e the use of *concepts*' (Ibid., p. 107).

Weberian 'logical analysis of concepts' stresses two types of relations between concepts, affinity and opposition. Such relations show a striking correspondence to the couple termed as syntagm and paradygm, the two axes of language, as applied to the study of culture. When Weber refers to the first type of relation in many passages of his work, the term he employs is 'elective affininity'. In its turn, the concept is quite close to Wittgenstein's 'family resemblance' which he uses in comparing language games[38] (and in opposing them to technically regulated practices[39]), both of them referring to the 'morphologi-

36 'The use of the indifferentiated collective concepts of everyday speech is always a cloak for confusion of thought and action. It is, indeed, very often an instrument of specious and fraudulent procedures. It is, on brief, always a means of obstructing the proper formulation of the problem' (Meth., p. 110).

37 Cf. Meth., pp. 68-69, 110.

38 'I can think of no better expression to characterize these similarities than "family resemblances"; for the various resemblances between members of a family: build, features, colour of eyes, gait, temperament, etc. etc. overlap and criss-cross in the same way. – And I shall say: "games" form a family' (PI, 67). See also PG, 35; PI, 65-71.

39 'Why don't I call cookery rules arbitrary, and why am I tempted to call the rules of grammar arbitrary? Because "cookery" is defined by its end, whereas "speaking" is not. That is why the use of language is in a certain sense autonomous, as cooking and washing are not. You cook badly if you are guided in your cooking by rules other than the right ones; but if you follow other rules than those of chess you are *playing another game*; and if you follow grammatical rules other than

cal model' which Goethe adopted in order to study natural phenomena[40]. In order to understand Weber's idea of history, and consequently the logic of his social science, no term is more crucial then 'elective affinity' *(Wahlverwandtschaft)*, R.H. Howe maintains in *Max Weber's Elective Affinities: Sociology Within the Bounds of Pure Reason* (1978). Instances of elective affinity in Weber's work are rare[41], his usage is diverse and moreover quite informal, but to discover what elective affinity meant to Weber is to discover 'its place in the order

such-and-such ones, that does not mean you say something wrong, no, you are speaking of something else' (Z, 320). Such distinction is in turn strictly related to Weber's distinction between the two ideal types of rationality, instrumental rationality and value-rationality.

40 Wittgenstein's adoption of a grammatical method in studying language phenomena is strictly related to the morphological method Goethe uses to study natural phenomena (Cf. Marconi 1996, Marconi 1997, Andronico 1997). According to Goethe the natural scientist is to develop a comparative analysis of the numberless forms natural beings show, by steadily comparing them with an originary model, type or phenomenon, itself drawn out of the natural world, but assumed in an ideal realm, just because of the touchstone function it is called to perform. Such study aims at estabilishing orderly series of natural phenomena that might make visible the whole of the relations of analogy and difference, i.e. family relations, tying them together. The method's success depends both on the naturalist's sensitivity and on his ability to fill in the blanks that may subsist among members of a series by inventing intermediate members that will make the passage from one form to another understandable. In contrast to Newton's mechanical method, Goethe rejects causal explantions and aims at *'darstellen und nicht erklären'* natural phenomena such as they appear to any observer's eye, without searching for anything hidden behind. Wittgenstein largely employs almost every element of Geothe's morphological model but he proves reluctant to admit it because he uses it 'in a totally new perspective' (Andronico 1997, p. 277), distancing himself from Geothe's illusion about being able to use morphology as a theoretical model to counter Newton's approach to natural phenomena. According to Wittgenstein morphological analysis 'can tell us nothing on natural phenomena beacause it deals with our concepts of them' (Ibid., p. 282). In the same perspective Weber maintains that 'the knowledge of social laws is not knowledge of social reality but is rather one of the various aids used by our minds for attaining this end', and 'knowledge of *cultural* events is inconceivable except on a basis of the *significance* which the concrete constellations of reality have for us in certain *individual* concrete situations' (Meth., p. 80).

41 Howe lists sixteen occurrences, cf. p. 367.

of his discourse', the 'latent structure of his thought' (Ibid., p. 367). The informality and diversity of Weber's use of elective affinity obscure, according to Howe, what he may have meant by the term, despite the significance given it in the 'Protestant Ethic'. Once, however, in *Economy and Society,* Weber ventures a few remarks offering an explanation. After rehearsing the difficulties inherent in making either causal or functionalist propositions concerning the interaction of economy and society, Weber states that:

> [...] something universal can, however, be asserted concerning the degree of elective affinity of concrete structural forms of social action with concrete economyc form, that is, concerning whether and how strongly they mutually favor one another's continuance or, conversely, hinder or exclude one another – are 'adequate' or 'inadequate' to one another (E&S, p. 341).

These brief remarks, Howe maintains,

> can only suggest that for Weber the sole significant universal propositions that social science could make would be propositions of elective affinity, which would be to say that the very possibility of social science would rest on the logic of elective affinity. But why this would be so and what that logic would be for materials as diverse as those encompassed by his own usage remains unstated, and nowhere else did he elaborate upon the term. But then, he had no need to. His order of discourse took care of all that (Howe 1978, p. 369).

Through a complex procedure meant to retrace the meanings of 'elective affinity', from the origin of the term in chemistry to its recurrence in philosophy, starting with Kant, and in literature, starting with Goethe[42], Howe succeeds in reconstructing a Weberian 'order of discourse', locating within it a precise connection between the order of sociological investigation and the order of ordinary language. In 'The Conflict in the German Literature of the Past Decade concerning the Character of the Social Constitution of the Ancient Germans', Weber

42 'From chemistry and Bergman would come the basic paradigm of elective affinity; from literature and Goethe, its application to the portrayal of social relationship; from philosopy and Kant, the art of divorce of the empirical from the rational and the affinity of all things in their possibility' (Howe 1978, p. 382).

uses the concept of 'inner affinity' rather than 'elective affinity'[43], connecting it, however, with the concept of 'adequacy', as he will do later on in the passage from *Economy and Society* mentioned above. But more important here for the logic of elective affinity in Weber's usage are, according to Howe, his reference to 'the analysis of constructs into judgments' and his 'denial of necessity in the relationships between those judgments'. The former would yield the 'framework' for the idea of elective affinity; the latter, its 'locus' in the order of his thought (Ibid., p. 378).

His friend and colleague Heinrich Richter seems to be the one who provided a starting point for Weber's 'framework', using the metaphor of the 'net of threads' which defines our knowledge as a 'system of judgments', our judgments as 'connections between the concepts' (the 'knots' of the net), and our concepts as 'the transition points of intersecting judgments'[44]. Where the judgments constituting two or more of its terms intersect, Howe says, a transition point between the terms is defined. Where two or more of those terms share a predicate in common, they are in that respect 'adequate' to one another, or they possess 'inner affinity' (Ibid., p. 379).

43 'When we construct a "stage of culture", then this thought construct, analyzed into judgments, means only that the individual appearences that we thereby assemble conceptually are "adequate" to one another, possess a certain measure of "inner affinity" *[innere Verwandtschaft]* [...] with one another, but never that they follow from one another with any kind of lawfulness. In other words, they are a conceptual means of portrayal, but not foundations for a deduction in accord with the infamous schema "All men are mortal; Caius is a man; therefore, he is mortal"' (Weber 1904, quoted by Howe, op. cit. p. 378).

44 '[...] the logical ideal of our knowledge consists of a complete system of judgments, whose subjects and predicates are constant, therefore defined concepts. Let us imagine this systematization of our knowledge accomplished in that direction. We could then compare the content of our knowledge with a net of threads in which the fixed knots represent the concepts; then threads, on the other hand, which go from one knot to the other are to delineate the connections between the concepts, that is, the judgments... Human thought [...] could never "envisage" this net in its totality, but could only traverse it in that it now forms concepts from out of the elements connected to one another, that is, the judgments, now analyzes these concepts back into judgments [...] just as the knots in the net consist only of threads, concepts are nothing else but the transition points of intersecting judgments.' (Rickert 1888, quoted by Howe, op. cit., p. 378-379).

But these inner affinities hold not only for scientific concepts, but, 'as any dictionary demonstrates', according to Howe, they also hold for the words themselves 'in the referential networks of their ordinary usage'. In Weber's view, our experience is 'absolutely no longer to be separated from [...] the words', and weberian scientific typologies are absolutely not unrelated to the 'natural affinities of ordinary language' (Ibid.). As Weber points out,

> The constructive concepts of sociology [...] are ideal-typical not only outwardly but also inwardly. In the great majority of cases, real action proceeds in muffled half-consciousness of its 'intended meaning'. The actor vaguely 'feels' it rather than knowing it [...] Actually effective, that is, fully conscious and clear meaningful action is in reality always only a limiting case [...]. But that is not to hinder sociology in forming its concepts through classification of the possible 'intended meaning', which is to say as if the action did in fact proceed in conscious orientation to meaning (E&S, p. 21).

If the concepts of sociology are the limit cases of the possible intended meanings of words, and if the words are 'inner affine' to one another, then 'the inner affinities of the words will be reflected at their limit in those concepts. Whether inner or elective, the affinities of the concepts will have their source in the meanings of the words' (Howe 1978, p. 379).

What kind of relationship exists between 'action' and 'words'? 'Action', says Weber, 'is [...] to mean a human behavior [...] is and insofar as the actor or actors attach a subjective meaning to it' (E&S, p. 4). From this definition, Howe argues,

> it follows at once that every meaning in the lexicon of the actors represents a possible action, and that, conversely, every action portrays the meaning in that lexicon to which it is referred. Thus, the world, both actual and possible, is a *mundus intelligibilis*. (Howe 1978, p. 379).

The elements of this intelligible world are not to be considered in their relationship to thruth, nor are subjectively intended meanings to be considered psychologically. They are to be considered in their relationship to a language community as a 'consensual community': to be understood or to act appropriately and without censure, 'the actor's understanding of the lexicon must be largely congruent with that of their language community overall'; and insofar as this situation is the case, 'the affinities of the words would have objective force, if not "validity"' (Ibid., 380). The lexicon of a language community 'defines the universe of possible actions of its members', and each point of intersection of this lexicon presents to the actor 'a transition point of choice of possible actions'. Sociology cannot determine which choice of possible action the actor will make in his traversals of his universe of meanings, the actor's choice of 'ultimate values' is beyond the ken of social science. Therefore, Weber denies necessity in the relationship constituting the conceptual constructs of his science, despite all adequacy or inner affinity. But this is the 'locus' of elective affinity in the order of Max Weber's thought:

> It is the freedom of the actors with respect to their choices of ultimate values that renders Weber's social science problematic within the Kantian bounds of his order of discourse, even as this freedom derives from that Kantian frame. The logic of elective affinity would provide a solution. Here a metaphor may be of use. Values are related to meanings as are the constellations to the stars. They are not their source and in no way could knowledge of meanings determine choice of values. For the elements of the *mundus intelligibilis* are multivalent. Their affinities are manifold and may be of three kinds. They may join the elements of the *mundus intelligibilis* to one another through the intersection of their meanings. They may exclude those elements from one another through a like intersection with change of sign. There may be no intersection, but through the former all meanings are joined, however indirectly. A meaning in total isolation from the rest is no meaning at all. The greater the number of positive inner affinities between two elements vis-à-vis the total possible number, the more strongly are the elements joined. That is their 'degree' of elective affinity. The higher the meaning, the greater would be its elective affinity for all the rest. The highest are values. But the lexicon of possible action, of meanings, is finite and the harmony of the values by no means assured. [...] Only as the logic of the interrelationships of networks of meanings, of possible actions, could elective affinity extend across the diversity of Weber's usage, from the elective affinity of concrete forms of social action with concrete economic forms to the elective affinity of similar ideals. [...] As a maxim for the conduct of scholarship, elective affinity would suggest the construction of ideal-types derived from the universe of ordinary language and the analysis of those types according to their judgments in order to trace in their elective affinities the actor's choice of possible actions. That yelds a logic for a sociology – *die verstehende* – which would 'in the con-

> sciousness of the narrow limits into which it is banned' (E&S, p. 8) vouchsafe its possibility [...] (Howe 1978, pp. 381-382).

For Weber, according to Howe, in the great tradition of German philological scholarship, the order of a language was 'the virtual order of a society'. History would be a logical chaos were it not for an order in the universe of the meanings to which actors orient their actions. That order is to be found in the elective affinities of words, the greater or lesser extent to which they possess inner affinity through the intersections of their meanings. It is this order in the universe of possible actions which makes weberian social science possible. The actor's choices of possible actions are given by the elective affinities of their universe of meanings. The order of the actual, the course of history and the structure of society, is to be read from this order, and the task of Weber's science 'is to portray its changing constellations' (Ibid., p. 383).

4 Meaning, identity, authority.

Weber's actor's lexicon is not only rich in elective affinities but also in 'irreconciliable death-struggles' between values[45]. Within it one finds not only values that 'join' one another, in Howe's words, but also values of opposing 'sign' or values that are mutually exclusive. Among these one can mention the opposition between 'personal' and 'impersonal' orders, which Weber uses to construct the three pure types of authority and also exemplifies in its typical propositional form:

45 '[...] a system of "values", be it ever so well-ordered, is unable to handle the situation's crucial issue. It is really a question not only of alternatives between values but of an irreconcilable death-struggle, like that between "God" and the "Devil". Between these, neither relativization nor compromise is possible. At least, not in the true sense' (Meth., pp. 17-18). The 'true sense' points to Wittgenstein's grammatical sense, that Weber also employs as a criterion to measure 'relativizations' and 'compromises' to be caught in the reality of action.

> In the case of legal authority, obedience is owed to the legally established impersonal order [...] In the case of traditional authority, obedience is owed to the *person* of the chief who occupies the traditionally sanctioned position of authority and who is (within its sphere) bound by tradition. But here the obligation of obedience is a matter of personal loyalty within the area of accustomed obligations. In the case of charismatic authority, it is the charismatically qualified leader as such who is obeyed by virtue of personal trust in his revelation, his heroism or his exemplary qualities as so far as they fall within the scope of the individual's belief in his charisma (E&S, p. 215-217). From a substantive point of view, every charismatic authority would have to subscribe to the proposition, 'It is written [...] but I say unto you [...]' The genuine prophet, like the genuine military leader and every true leader in this sense, preaches, creates, or demands *new* obligations [...] (Ibid., p. 243).

In Weber's radically relational perspective, authority is no personal attribute to Ego, such as its 'soul' or 'atmosphere', but is rather made up by its initial 'claim to legitimacy', that grows into real authority only in case it is followed by some kind of 'belief in legitimacy', or basis for some 'motive of compliance' on the part of its necessary Other. In other words, Weber's actor's authority always takes shape within some 'language-game', and the language-game is founded, in this case, on the grammatical opposition between personal and impersonal orders. We are dealing here with two relatively necessary values, that is, values that are mutually exclusive *(aut-aut)*, like 'God' and 'Devil', rather than two relatively indifferent means *(vel-vel)*, that are each conversely relative to an external end. Weberian presentation of authority can be termed 'perspicuous' in Wittgenstein's sense, if we confer on the second type of authority, the traditional one, also the character of *'intermediate case'* meant to produce just that understanding which consists in 'seeing connections'[46] between the two opposing poles of his 'bipolar structure'[47], the legal and the charismatic ones.

46 Cf. n. 19.

47 Separating 'personal' from 'impersonal' provides the 'key opposition' (Rebuffa 1991, p. 49) or the 'bipolar structure' (Hennis 1991, p. 127) within Weber's more famous ideal type construction, i.e. the three types of authority. From a grammatical perspective, every 'claim to legitimacy' becomes a move within a game in the true sense of the word. By personally opposing the impersonal quality of writing – 'It is written... but I say unto you...' – one moves within a game of authority,

The same opposition between personal and impersonal values is to be found as the pivotal focus in Weber's lecture on science, *Science as a Vocation*, the text where weberian analysis of language and institutions, of identity and power, of the real and the imaginary, all close in and merge in his most successful literary manner, giving a token of what kind of rule and what kind of method Weber bequeathts to sociology. His presenting the actor's grammar of linguistic and cultural motives entails here the latter's presenting his own personal identity and his own claim to authority as well, while his analysis of language reality entails an anlysis of the deep-rooted mythology active within it, a whole mythology made of idols and monsters. I am now thinking of those passages in which Weber presents the diverse motives belonging to 'the specialist' and to 'the dilettante', to 'the teacher' and to 'the leader', and specifies the dilettante's idols and the monster produced by arbitrarily mingling teacher and leader. In this next step I will try to establish a new parallel between Weber's sociology and the philosophy of language, in the form of a parallel between Weber and J.R. Searle.

Wissenschaft als Beruf highlights the problem of the sense *(Sinnproblem)* of science, that is to say the problem science poses to whoever addresses it with questions having to do with meaning. Weber solves this problem in a grammatical way, by presenting a variety of typical motives belonging to the ones who work within science, starting from the reality of scientific organization, as it has been developing in the course of time. In our time, Weber says, science has entered a phase of specialization 'previously unknown', and this 'will forever remain the case'. Not only outwardly, but inwardly, matters

---

made up by the presence of those two values. Every claim to legitimacy is an empirical possibility the meaning of which is given by a grammatical necessity: no charismatic or personal authority can assert itself without a legal or impersonal authority. Weberian ideal types of authority, taken in the sequence Weber presents them in (from legal authority to charismatic authority) do not 'invert' historical order (Ricoeur 1994, p. 232) because, as ideal types, they do not 'describe' history but rather locate a linguistic and cultural order within which personal and impersonal values are necessarily tied to each other and as such work as standards of measurement for history seen as a product of the social actor as a 'cultural being'. They allow us to describe history just because they oppose any myth on 'origins'.

stand at a point where the individual can acquire the sure consciousness of achieving something truly perfect in the field of science only in case he is a 'strict specialist' (FMW, p. 134). A really definitive and good accomplishment is today always 'a specialized accomplishment', and whoever lacks the capacity 'to put on blinders' and 'to come up to the idea that the fate of his soul depends upon whether or not he makes the correct conjecture at the passage of this manuscript' may as well stay away from science: he will never have the 'personal experience' of science (Ibid., p. 135). Now, while this very impersonal and at the same time passionate way to come about a personal experience of science turns a scientist into a specialist, the same does not hold true for the dilettante who, taking apart the components of that whole, turns his own personal experience into an idol to be worshipped. The dilettante differs from the specialist not because the latter only may produce 'ideas', but because he proves the only one to be 'in the position to control, to estimate, or to exploit the idea in its bearing'. The specialist can reach that position only in so far as he impersonally serves the subject of his work, whereas the dilettante pretends to use this subject with the single aim to replace it himslef in his own person.

> Now, whether we have scientific inspiration depends upon destinies that are hidden from us, and besides upon 'gifts'. Last but not least, because of this indubitable truth, a very understandable attitude has become popular especially among youth, and has put them in the service of idols whose cult today occupies a broad place on all street corners and in all periodicals. These idols are 'personality' and 'personal experience'. Both are intimately connected, the notion prevails that the latter constitutes the former and belongs to it. People belabor themselves in trying to 'experience' life – for that befits a personality, conscious of its rank and station. And if we do not succeed in 'experiencing' life, we must at least pretend to have his gift of grace [...] In the field of science only he who is devoted *solely* to the work at hand has 'personality'. And this holds not only for the field of science [...] In the field of science, however, the man who makes himself the impresario of the subject to which he should be devoted, and steps upon the stage and seeks to legitimate himself through 'experience', asking: How can I prove that I am something other than a mere 'specialist' and how can I manage to say something in form or in content that nobody else has ever said? – such a man is no 'personality'. Today such conduct is a crowd phenomenon, and it always makes a petty impression and debases the one who is thus concerned. Instead of this, an inner devotion

to the task, and that alone, should lift the scientist to the height and dignity of the subject he pretends to serve (Ibid., p. 137).

If an idol is brought about by separating an image from the whole every image is part of, personality becomes an idol when, as is the case with the dilettante, it is separated from an impersonal practice such as the specialist's 'inner devotion to the task', whose personality is far from being only imaginary. Once 'personal experience' too is assumed by the dilettante as an idol, and Weber insists on it[48], not even a connection between two idols, no matter how inward or innermost, will bring about anything but idols: not more than, in Wittgenstein's terms, buying 'several copies of the morning paper' may garantee that 'what it said was true'[49], or than considering a gift the money put 'into (one's) left hand' by '(one's) right hand'[50].

In further passages, Weber presents his public with the criteria he uses to identify the 'true' master and the 'true' leader as two grammatically opposed identities, whose overlapping may only cause a monster in the true sense of the word. As he does so, he keeps steadily together what in fact is really always one, that is to say, the actor's motive's linguistic text and the cultural context of its occurrence. The monster, in its mythological form, is the result of mingling two distinct and incompatible natural forms, as is the case in the mythological figure of the Mermaid. As he basically distinguishes between the two different practices of scientific investigation and value judgements, as he does throughout his writings, Weber finds them at work within the two forms of life of teaching and politics assuming them as two given realities, having nothing in common, each of them showing, on the

48 See also *The meaning of 'Ethical Neutrality' in Sociology and Economics*: 'The powerful personality does not manifest itself by trying to give everything a "personal touch" at every possible opportunity. The generation which is now growing up should, above all, again become used to the thougt that "being a personality" is something that cannot be deliberately striven for and that there is only one way by which it can (perhaps!) be achieved: namely, the whole-hearted devotion to a "task" whatever it (and its derivative "demands of the hour") may be' (Meth., p. 5).

49 PI, 265.

50 PI, 268.

contrary, opposite traits and each of them being characterized by incompatible truth criteria:

> To take a practical political stand is one thing, and to analyze political structures and party positions is another. When speaking in a political meeting about democracy, one does not hide one's personal standpoint; indeed, to come out clearly and take a stand is one's damned duty. The words one uses in such a meeting are not means of scientific analysis but means of canvassing votes and winning over others. They are not plow-shares to loosen the soil of contemplative thought; they are swords against the enemies: such words are weapons. It would be an outrage, however, to use words in this fashion in a lecture or in the lecture-room [...] But the true teacher will beware of imposing from the platform any political position upon the student, whether it is expressed or suggested [...] Now one cannot demonstrate scientifically what the duty of an academic teacher is. One can only demand of the teacher that he have the intellectual integrity to see that it is one thing to state fact, to determine mathematical or logical relations or the internal structure of cultural values, while it is another thing to answer questions of the *value* of culture and its individual contents and the question of how one should act in the cultural community and in political association. These are quite heterogeneous problems. If he asks further why he should not deal with both types of problems in the lecture-room, the answer is: because the prophet and the demagogue do not belong on the academic platform. [...] Those of our youth are in error who react to all this by sayng, 'Yes, but we happen to come to lectures in order to experience something more than mere analyses and statements of fact.' The error is that they seek in the professor something different from what stands before them. They crave a leader and not a teacher. But we are placed upon the platform solely as teachers. And these are two different things, as one can readily see (FMW, p. 145-146, 149).

Weber's social actor finds his substance in language as an unchanging factor[51], by taking on a changing identity, depending on the form of life within which he speaks. By locating an actor's motivational texts within the contexts in which they were brought forth, Weber stresses his cultural nature, and lays the foundations for a kind of realism which runs counter to both the subjective strain of private language[52]

51 For a comparison between Weber's social 'actor' and Benveniste's linguistic 'person' see A. Sormano (2000).

52 '[...] the ability to perform a similar action is not a necessary prerequisite to understanding; "one need not have been Caesar in order to understand Caesar"' (E&S, p. 5).

and the objective strain of indifferentiated collective concepts[53]. He does so by assuming cultural and linguistic institutions as criteria of intelligibility, that is to say as data. Within the bounds of this kind of 'realism', the very coming up of totally imaginary idols and monsters that is so widespread in every culture, becomes linguistically and culturally intelligible, and therefore liable to be described.

By drawing a line between the professor's role identity and the teacher's pragmatic identity and differentiating descriptive from value oriented language games, Weber recognizes the teacher and the leader as personal identities that are made up by using language in different ways. The rule of keeping free from value judgements lies at the basis of any scientific practice or institution and every professor is bound to follow it if he wants to present himself as a teacher within the language of 'fact descriptions' and of 'analysis', within the institutional context provided by the lecture room. On the contrary, value judgements lie at the basis of political institutions, and he who wants to be a leader within the quite different institutional context provided by, say, a political meeting, is to express himself through value judgements. In the first case, the teacher will describe reality, that is, the subject of his teaching, in terms that can be metaphorically called 'plow-shares to loosen the soil of contemplative thought'; in the second case, his taking his stand as to the various values will be 'swords' the leader will use 'against the enemies'. Using words as weapons within the political context is a leader's 'damned duty', but it becomes 'an outrage' if it is used in a university context as a surplus to actual analysis and scientific descriptions[54]. Such an outrage will turn the professor into a two-

53 'For still other cognitive purposes – for instance, juristic ones – or for practical ends, it may on the other hand be convenient or even indispensable to treat social collectivities, such as states, associations, business corporations, foundations, as if they were individual persons. Thus they may be treated as the subjects of rights and duties or as the performers of legally significant actions. But for the subjective interpretation of action in sociological work these collectivities must be treated as *solely* the resultants and modes of organization of the particular acts of individual persons, since these alone can be treated as agents in a course of subjectively understandable action' (E&S, p. 13).

54 See also *The meaning of 'Ethical Neutrality' in Sociology and Politics*: 'Like everyone else, the professor has other facilities for the diffusion of his ideals. When

headed monster bearing both the teacher and leader's head, rather than allowing him to be a real, full-fledged teacher. He will act as a Mermaid whose singing will probably allure many students.

The two kinds of linguistic games are irretrievably opposed to each other because they belong to diverse forms of life having each its own truth criteria. Anticipating Wittgenstein's formulations, Weber assumes science and politics as two forms of life to be accepted as 'the given'[55]; two forms whose rise and empirical existence are no doubt arbitrary, but once they have been established, having an existence that becomes a necessary criterion to understand what comes up within its bounds. This kind of necessity belongs to any grammar, or any institution as far as they are the basis of a grammatical order. Orders are at times disobeyed, Wittgensteins says, 'but what would it look like if orders were *never* obeyed? The very notion of order would have lost its purpose' (PI, 345). As Weber says, in a lecture-room at times a professor takes his stand or pretends he can claim scientific foundation for a value judgement whatsoever; but what would it look like if *every* professor took their stand as to values instead of describing them technically? The very notion of university would have lost

these facilities are lacking, he can easily create them in an appropriate form, as experience has shown in the case of every honest attempt. But the professor should not demand the right as a professor to carry the marshal's baton of the statesman or reformer in his knapsack. This is just what he does when he uses the unassailability of the academic chair for the expression of political (or cultural-political) evaluations' (Meth., p. 5).

55 'What has to be accepted, the given, is – so one could say – *forms of life*' (PI, 226). Philosophical (and sociological) elucidation of 'human intelligence, and the notions associated with this', according to Winch, 'requires that these notions be placed in the context of the relations between men in society', i.e. in Wittgenstein's 'forms of life'. To understand what is involved in a 'form of life' is the aim not only of epistemology, according to Winch, but also of the 'theoretical part of general sociology': 'It is true that the epistemologist's starting point is rather different from that of the sociologist but, if Wittgenstein's arguments are sound, that is what he must sooner or later concern himself with. That means that the relations between sociology and epistemology must be different from, and very much closer than, what is usually imagined to be the case.' (Winch 1958, 40-42) Winch's encouragement seems to have been anticipated by Weber, although Winch recognizes it only in part (Sormano 2000).

its purpose, i.e. the university as an institution would cease to exist, would lose its place within grammar, would be no longer *so* if compared to its *otherwise*, such as, for instance, a party meeting. As a social scientist, Weber does not defend institutions, nor does he accept them as collective concepts having their own built-in validity. In his individualistic descriptive activity he rather has recourse to them steadily as a criterion meant to understand social actions, as an internal element within his method of observing an actor's motives[56]. Describing such motives goes along with describing the personal identity of the one who expresses them, but also runs parallel to recognizing the kind of authority the actor enacts in that way, as we shall soon see. In locating the contextual meaning of motives of action, Weber recognizes at the same time, and through the same criterion, the actor's personal identity and claim to power. If we turn to J.R. Searle for help, we may say that Weber recognizes and describes the rise of 'a large number of institutional facts', within the language pertaining to each motive.

Institutional facts, Searle argues in *The Construction of Social Reality* (1995), are inseparable from 'brute facts', logically coming before them, and are constituted by language according to the typical form: 'X counts as Y' or 'X counts as Y in context C'. The move from the brute (X) to the institutional status (Y) is '*eo ipso* a linguistic move, because the X term now symbolizes something beyond itself' (Ibid., p. 73). Among other instances, Searle mentions marriage:

> Consider, for example, the sort of speech act people perform in a marriage ceremony. Performing such and such speech acts (the term X) in front of a presiding official (the term C) now counts as getting married (the term Y). Saying those very same words in a different context, while making love, for

56 The passage quoted in note 51 continues thus: 'Nevertheless, the sociologist cannot for his purposes afford to ignore these collective concepts derived from other disciplines. [...] These concepts of collective entities which are found both in common sense and in juristic and other technical forms of thought, have a meaning in the minds of individual persons, partly as of something actually existing, partly as something with normative authority. This is true not only of judges and officials, but of ordinary private individuals as well. Actors thus in part orient their action to them, and in this role such ideas have a powerful, often a decisive, causal influence on the course of action of real individuals' (E&S, p. 13-14).

> example, will not constitute getting married. The Y term now assigns a new status to those speech acts. [...] If we are right in thinking that marriage is typical of many institutions, then it is a consequence of the account that there will be a hierarchical structure of creation of a large number of institutional facts. Thus, going through the example of marriage: Making certain noises counts as uttering an English sentence, uttering a certain sort of English sentence in certain circumstances counts as making a promise, making a promise in certain circumstances counts as entering into a contract, entering into certain sorts of contracts counts as getting married. The marriage ceremony creates a new institutional fact, the marriage, by imposing a special function on a set of speech acts. But the creation of the marriage imposes a new status and therefore a new function on the individuals involved. They are now 'husband' and 'wife'. And the fact that they are husband and wife, like the marriage itself, is an istitutional fact (Searle 1995, pp. 82-83).

Once they are defined in this way, status relationships become at the same time power relationships. 'The structure of institutional facts is a structure of power relations, including negative and positive, conditional and categorical, collective and individual power'. Even if the whole idea of power makes a certain type of liberal sensibility 'very nervous', institutional power relations 'are ubiquitous and essential', institutional power – massive, pervasive, and typically invisible – 'permeates every nook and cranny of our social lives, and as such it is not a threat to liberal values but rather the precondition of their existence' (Ibid., p. 94)

There is no problem in translating Weber's passages in his lecture on science employing Searle's rule:

1. making certain noises (X) counts as analyzing (Y) in our language (C)
2. analyzing (X) counts as 'holding a lecture' (Y) in a lecture-room (C)
3. holding a lecture (X) in a lecture-room (C) counts as being a 'teacher' (Y)

Description is an institutional fact within language (1). When it takes shape within a cultural context such as a university, it gives way to a new istitutional fact, a lecture (2) which in turn bestows new status functions and corresponding powers on the individuals involved in such activities, that is professors and their students (3). The same criterion identifies the various status functions and corresponding powers concerning leaders and their followers, starting from taking a stand, itself another insititution working within language.

1. making certain noises (X) counts as 'taking a stand' (Y) within our language (C)
2. taking a stand (X) counts as being involved in politics (Y) in a political meeting (C)
3. being involved in politics (X) in a political meeting (C) counts as being a 'leader' (Y)

Truth criteria applying to the 'true' teacher and the 'true' leader are constituted by the twofold institutional order, both cultural and linguistic, the necessary connection of which Weber is now showing us. The very criterion of *culturally contextualizing the linguistic texts in which a social actor's motives come to light*, (see below), allows Weber, in the same lecture, to point out the false teacher and in the following lecture on *Politics as a Vocation*, the false bureaucrat. They are two diverse realities the truth of which lies in their being two hybrids, or institutional monsters, brought about by mingling contradictory and conflicting texts and contexts.

TEXTS \ CONTEXTS	University	Politics
Value free (technichal analysis)	Teacher	Techno-crat
Value oriented (value-judgments)	Teach-leader	Leader

The teacher. Certain words brought together in a certain way, (as for instance 'the internal structure of cultural values is so and so') count, firstly, as technical analysis in the cultural context our language supplies, the rules of which we have learned in the course of our socializing and that allow us to tell a description from an order, a promise from an encouragement, a wish from a charge and so on, even if we do not have a theory of speech acts at hand. Provided that such words be phrased according to game rules, we understand them 'immedately', according to Weber[57], 'in a flash' in Wittgenstein's words[58], without having to 'interpret' them by going through the psychic processes belonging to the one who utters them[59]: we immediately understand their literal meaning, telling it apart from other literal meanings that may be present within that same institution. If we move within a language game we also move within some other game, institution or form of life, whatever we may want to call it. The motives for action provide the subject for a science of reality. Its reality is based first of all on the fact that 'the speaking of language', in Weber's view, and later on in Wittgenstein's, 'is part of an activity, or a form of life', and on the assumption that such form of life – the 'given', what 'has to be accepted' – is an essential element in understanding the individual meaning of what is spoken within it, that is to say, an actor's real motives.

In Weber's view, however, discovering the individual or contextual meaning of the actor's motives goes along with discovering a personal identity and a kind of power. If the above mentioned technical analysis – 'the internal structure of cultural values is so and so' – is formulated within a university class-room, its linguistic act does not contradict the rule of ethical neutrality on which a university insitution is based, but rather confirms it, while setting it apart from other institutions. Such analysis acts as individual legitimation of the institution's claim to impersonal power, that is enforcing the rule of ethical neutrality. In this way, both rule and claim turn 'real'. By sanctioning claims to impersonal power one also establishes one's identity: the professor performing it presents himself as a 'true teacher' according to the truth criteria functioning within that context. By accepting such sanction and not expecting the teacher to prove a leader too, one will become in

57 Cf. E&S, p. 8 and n. 7.

58 Cf. PI, 138.

59 For Weber's criticism of Simmel's distinction between subjectively intended and objectively valid meanings see R&K.

his turn a true student. Nothing personal will then connect teachers and students; on the contrary, the impersonal legal authority of a university institution will prevail. The difference between a university and a party meeting lies in value judgements being at home only in the second kind of context. Only where parliaments, or monarchs, or revolutionary power-holders intervene for political reasons in academic selections, states Weber in the same conference, 'can one be certain that convenient mediocrities or strainers will have the opportunities all to themselves' (FMW, p. 133). Chances are that in these cases the impersonal rule of keeping free of value judgements will not be sanctioned. Lecturing in a lecture room through analysis, descriptions, explanations, instead of taking a stand or pretending to scientifically found value judgements, is itself a possibility the psychological reasons of which are totally irrelevant to describing its sense, when we realize that this possibility occurs in the form of a text, motive, or vocabulary, within that context. Assuming the rules of such text and context in their necessary connection is a prerequisite for recognizing and describing the real meaning of the subject one is enquiring into, as well as the kind of identity and power which come into being in any language occurrence.

The leader. The rule we followed in identifying the meaning of the true teacher's actions holds true in the case of the true leader. We immediately allow that a word group such as 'an X value is higher than a Y value' is different from the group 'the internal structure of cultural values is so and so': the former is a value judgement while the latter is a technical analysis. To take a practical political stand in a political meeting is a politician's 'damned duty', an act by which the politician asserts himself as leader while at the same time sanctioning the value rules of the political game.

The teach-leader. We are now considering an identity conferring rule belonging to Weber's science of reality. Such reality however is also made up by actions based on merely imaginary representations of reality. Weber introduces such representations together with their own kinds of indentity and power on the basis of their rule. In the above layout I used teach-leader in order to point to somebody who overlaps in an imaginary way forms of behaviour which are usually taken separately in the public realm of institutions, giving way to a so called monster, and, technically speaking, becoming a teach-leader. Weber has in mind anybody who might claim to found value judgements on science within a university insititution which by itself can only allow analysis and description[60]. If to take a practical political stand is a politician's damned duty, a teacher doing it is 'an outrage', i.e. it means desacrating a place devoted solely to the scientist's impersonal value free activity which results into an act delegitimizing that very impersonal order. Such act implies that somebody is claiming a personal power, based on a principle which equals the one the charismatic leader bases his own claim to power on: 'It is written [...] but I say unto you [...]'.

The techno-crat. Elsewhere, Weber describes the technocrat applying the same criterion referred to above. The technocrat is he who, while acting as a politician, does not accept the responsibility for his own stand, or he who while acting as an official, does not restrain himself to following the politician's directions. In his essay on 'objectivity', Weber deals with the former instance, considering a case of 'illusion of the self-evidence'. This takes place when, in a poltical context, one

60 In his essay on 'objectivity' Weber had already mentioned similar cases of 'confusion', together with their linguistic formulation: 'on the one hand (analysis of reality) [...] and on the other (value judgments of reality)'. '[...] the modern relativistically educated historian who on the one hand seeks to "understand" the epoch of which he speaks "in its own terms", and on the other still seeks to "judge" it, feels the need to derive the standards for his judgments from the subject-matter itself, i.e., to allow the "idea" in the sense of the *ideal* to emerge from the "idea" in the sense of the "ideal-type". The esthetic satisfaction produced by such a procedure constantly tempts him to disregard the line where these two ideal types diverge [...] In contrast with this, the *elementary duty of scientific self-control* and the only way to avoid serious and foolish blunders requires a sharp, precise distinction between the logically *comparative* analysis of reality by ideal-*types* in the logical sense and the *value-judgments* of reality *on the basis of ideals*' (Meth., p. 98). A similar position is taken in *The meaning of 'Ethical Neutrality' in Sociology and Economics*: 'Nothing is ever gained in any scientific sense whatever by "on the one hand", and "on the other", by seven reasons "for" and six "against" a certain event [...]' (Ibid., p. 24).

only argues about the best means to reach goals rather than disputing on the cultural values on which they are based.

> The distinctive characteristic of a problem of a social *policy* is indeed the fact that it cannot be resolved merely on the basis of purely technical considerations which assume already settled ends. Normative standards of value can and must be the object of *dispute* in a discussion of a problem of social policy because the problem lies in the domain of general *cultural* values (Meth., p. 56).

In his lecture, *Politics as a Vocation*, Weber defines the identity of the 'the genuine official' in a way that mirrors the one he used in his lecture on science:

> According to his proper vocation, the genuine official [...] will not engage in politics. Rather, he should engage in impartial 'administration' [...] *Sine ira et studio*, 'without scorn and bias' he shall administer his office. Hence, he shall not do precisely what the politician, the leader as well as his following, must always and necessarily do, namely, *fight*. [...] The honor of the civil servant is vested in his ability to execute conscientiously the order of the superior authorities, exactly as if the order agreed with his own convinction. [...] Without this moral discipline and self-denial, in the highest sense, the whole apparatus would fall to pieces (FMW, p. 95).

If an official does not impersonally execute his superior's order, he plays 'another game' rather than the administration's one, suspending the ethical discipline which allows its functioning on impersonal, technical rules. In other words, he turns the administration into an improper form of politics which may result into the whole apparatus crumbling down insofar as it is founded on that proper distinction.

It has been observed that Weber almost always uses the adjective bureaucratic and the noun bureaucracy giving them a positive value[61]. When Weber forsees the dangers involved in bureaucratization he imputes them to the mingling of the politician's and the official's roles, just as he did in the case of the university. The dangers do not arise when the politician performs his damned duty in taking a stand in the name of such and such value, they rather arise when somebody counteracts such confrontational game by entering an imaginary 'meta-

61 Cf. Rebuffa (1991, p. 61).

game' which aims at achieving a 'conservative mythology'[62] according to which, in Weber's times, bureaucracy was held to be an unbiased, universal entity, while today we would define it a technocracy. Weber may have feared that kind of meta-game based on somebody claiming to be beyond partial involvement, therefore denying any necessary confrontation when one deals with values, in the name of an absolute technical rationality. In so doing one also denies that politics is based on conflicts. It has also been observed that Weber cannot be termed a proper 'liberal' because conflict is a key concept in his thought and such concept represents the keenest retort against the liberal idealized view of peaceful competition, the so called advantages of the division of labour, and the euphemistic conceiving of a common interest for everyone, every class and every nation[63].

On the basis of my reading of Weber's analysis of an actor's motives, I would like to suggest that the basic tenet of liberal thought, i.e. the cultural representation of the division of powers, is filtered by Weber through a grammatical analysis. In other words, by analyzing grammatically, Weber can locate and criticize the overlapping of different powers, as opposed to their separation, seen as one form of mythology among others, that can be discried first of all by analysing language. If this is so, we can assume Weber's individualistic method as a sociological grammar, that is to say as a method we can profitably use to analyse meanings within language in action. By this I mean a method we can employ to locate both contextual meanings within the speaking of language and, at the same time, an actor's local identity, along with the type of claim to power he aims at in speaking. These three research subjects are often considered apart from one another, while the novelty of Weber's perspective lies exactly in his recognizing their necessary connection that gives shape to the real sense of social action whether it be actual or imaginary. The connection between meaning, identity and power assumed as an inseparable unit, working towards establishing the sense of social actions, whether it be a real or imaginary one, as rooted deeply in language, is a sociological fall-out which the sociologist looking for new sources and foundations may

62 Cf. Beetham (1989, p. 95).
63 Cf. Hennis (1991, p. 240).

gain today from reading Wittgenstein's philosopy as a chance for re-reading Weber's sociology.

Bibliography

M. Albrow, *Max Weber's Construction of Social Theory,* Mamillan, London 1990.

M. Andronico, *Giochi linguistici e forme di vita,* in D. Marconi (ed.), *Wittgenstein*, Laterza, Bari 1997.

D. Beetham, *La teoria politica di Max Weber*, Il Mulino, Bologna 1989.

A. Blum and P. McHugh, 'The Social Ascription of Motives', *American Sociological Review*, 36, 1971, pp. 98-109.

R.H. Brown, 'Rhetoric, Textuality, and Postmodern Turn in Sociological Theory', *Sociological Theory*, 8, 1990, pp. 188-197.

P.S. Cohen, 'Models', *The British Journal of Sociology*, 17, 1966, pp. 70-78.

J. Coulter, *Mind in Action*, Polity Press, Cambridge 1989.

W. Hennis, *Il problema Max Weber*, Laterza, Bari 1991.

R.H. Howe, 'Max Weber's Elective Affinities: Sociology Within the Bounds of Pure Reason', *American Journal of Sociology*, 84, 1978, pp. 366-385.

D. Käsler, *Max Weber. An Introduction to his Life and Work*, Polity Press, Cambridge 1988 [1979].

D. Marconi, *Wittgenstein e la filosofia*, in L. Wittgenstein, *Filosofia*, Donzelli, Roma 1996.

D. Marconi (ed.), *Wittgenstein*, Laterza, Bari 1997.

D.W. Maynard, 'Language and Social Life', *Contemporary Sociology*, 20, 1991, pp. 841-843.

C.W. Mills, 'Language, Logic, and Culture', *American Sociological Rewiew*, 5, 1939, pp. 670-680.

C.W. Mills, 'Situated Actions and Vocabularies of Motive', *American Sociological Review*, 5, 1940, pp. 904-913.

B.T. Pentland, 'Organizing Moves in Software Support Hot Lines', *Administrative Science Quarterly*, 37, 1992, pp. 527-548.

B.T. Pentland, H.H. Rueter, 'Organizational Routines as Grammars of action', *Administrative Science Quarterly*, 39, 1994, pp. 484-510.

G. Rebuffa, *Nel crepuscolo della democrazia. Max Weber tra sociologia del diritto e sociologia dello Stato*, Il Mulino, Bologna 1991.

P. Ricoeur, *Conferenze su ideologia e utopia*, Jaca Book, Milano 1994.

G. Roth, 'Socio-Historical Model and Developmental Theory: Charismatic Community, Charisma of Reason and the Counterculture', *American Sociological Review*, 2, 1975, pp. 148-157.

G. Roth, 'History and Sociology in the Work of Max Weber', *The British Journal of Sociology*, 27, 1976, pp. 306-318.

M.B. Scott and S.M. Lyman, 'Accounts', *American Sociological Review*, 33, 1968, pp. 46-62.

J.R. Searle, *The Construction of Social Reality*, Allen Lane, The Pinguin Press, London 1995.

A. Sormano, *Grammatica del senso. Weber, Wittgenstein, Benveniste*, Libreria Stampatori, Torino 2000.

A. Sormano, *Sociologia e linguaggio*, Gruppo Editoriale Esselibri, Simone, Napoli 2003.

F. Tessitore, *Lo storicismo*, Laterza, Bari 1996.

J.W.N. Watkins, 'Ideal Types and Historical Explanation', *British Journal for the Philosopy of Science*, 3, 1952, pp. 22-43.

M. Weber, FMW, *From Max Weber: Essays in Sociology* (translated, edited, and with an introduction by H.H. Gerth and C.W. Mills), Oxford University Press, Oxford 1946.

M. Weber, Meth., *The Methodology of the Social Sciences*, (translated and edited by E.A. Shils and H.A. Finch), Glencoe, Illinois 1949.

M. Weber, R&K, *Roscher and Knies: The Logical Problems of Historical Economics*, Free Press, New York 1975.

M. Weber, E&S, *Economy and Society*, (edited by G. Roth and C. Wittich, 1st printing 1968), University of California Press, Berkeley 1978.

P. Winch, *The Idea of a Social Science and its Relation to Philosophy*, Routledge & Kegan Paul, London 1958.

L. Wittgenstein, PI, *Philosophical Investigations* (translated by G.E.M. Anscombe), Basil Blackwell, Oxford 1953.

L. Wittgenstein, Z, *Zettel*, (edited by G.E.M. Anscombe and G.H. von Wright, translated by G.E.M. Anscombe) Basil Blackwell, Oxford 1967.

L. Wittgenstein, P, *Philosophy*, in J. Hintikka (ed.), *Wittgenstein in Florida*, Kluwer Academic Publishers, Dordrecht 1991.

L. Wittgenstein, PG, *Philosopical Grammar* (edited by R. Rheed, translated by A. Kenny), Basil Blackwell, Oxford 1974.

L. Wittgenstein, RFGB, *Remarks on Frazer's Golden Bough* (translated by J. Beversluis), in C.G. Luckhard (ed.), *Wittgenstein. Sources and Perspectives*, The Harvester Press, Hassocks, Sussex 1979.

L. Wittgenstein, *Filosofia*, (edited by D. Marconi, translated by M. Andronico), Donzelli, Roma 1996.

Andrea Sormano, Dipartimento di Scienze Sociali, Università di Torino

M.U.S.A. (Models of the User in Advanced Societies)

A nationwide project supported by the Italian Ministry of University Teaching and Research (M.I.U.R.): *Models of the User in Advanced Societies.*

National Coordinator: Prof. Massimo Negrotti, University of Urbino, Italy.

Local Coordinators:

Danila Andreina Bertasio, dan@soc.uniurb.it
The user of the art between public and private.

Maria Domenica Combi, mariella.combi@uniroma1.it
The multicultural user and new technologies. A cultural anthropological approach.

Franco Fileni, filenif@sp.univ.trieste.it
Internet: the user between usability and culture. The case of World Wide Web.

Renata Livraghi, renata.livraghi@unipr.it
Quality in University as effect of good curricula *and strategic interaction to users.*

Giuseppe Longo, longo@univ.trieste.it
The concept of user with reference to the information and communication technologies.

Massimo Negrotti, massimo.negrotti@libero.it
The user of the artificial: methodological problems and cultural issues.

Giuseppe Padovani, giuseppe.padovani@unipr.it
The user in public communication.

Antonino Porrello, porrello@iuav.unive.it
The user of advanced technologies: the user, the artificial and the simulation technologies.

Raimondo Strassoldo di Graffembergo, raimondo.strassoldo@dest.uniud.it
The user of the art services.

Mariselda Tessarolo, mariselda.tessarolo@unipd.it
Organizational experience of cultural users.

Massimo Negrotti

Naturoids: On the Nature of the Artificial

Since antiquity, technology has tried to either control or imitate nature. Both these traditions take advantage of the progress of science, but their teleology and their typical design problems remain basically different.

The technology of the artificial may be defined as the effort to reproduce natural objects or processes by means of current conventional technology and materials. This book reports on the results of a theoretical study of the logic characterizing any attempt to design something artificial.

While designers of artificial devices work in their own area, facing field-specific problems (for example, bioengineering, artificial organs, robotics, AI, ALife), the present study refers to the artificial in itself, trying to discover what is common to instances very distant from each other, in an intrinsically interdisciplinary way. The result may be defined as a proposal of a general theory of *naturoids*.

World Scientific, New Jersey, 2002

Yearbook of the Artificial

Published volumes

Vol. 1 Yearbook of the Artificial: Nature, Culture & Technology, 272 pages, 2002.

Vol. 2 Yearbook of the Artificial: Models in Contemporary Sciences, 379 pages, 2004.

FSC
www.fsc.org
MIX
Papier | Fördert gute Waldnutzung
FSC® C083411

Kolibri 360 GmbH
Ferdinand-Jühlke-Straße 7
99095 Erfurt, Deutschland
produktsicherheit@kolibri360.de

Druck:
CPI Druckdienstleistungen GmbH
im Auftrag der
Kolibri 360 GmbH
Ferdinand-Jühlke-Str. 7
99095 Erfurt